Hamlyn Colour Guides

Minerals

Hamlyn Colour Guides

Minerals

by Jaroslav Švenek

Illustrated by
Ladislav Pros

HAMLYN

Translated by Zdena Náglová
Graphic design by František Prokeš
Designed and produced by Artia for
The Hamlyn Publishing Group Limited,
Bridge House, London Road, Twickenham,
Middlesex, England
ISBN 0 600 30657 7
Printed in Czechoslovakia by Svoboda, Prague
3/15/19/51-01

CONTENTS

MINERALS AND MAN

Since prehistoric times the mineral world has attracted the attention of man, from the practical point of view as well as with its richness of colour and the beauty of crystals. The importance of minerals as raw materials in the evolution of mankind is best expressed by the division of the early stages of man's existence into the Stone, Bronze, and Iron Ages. Many minerals used by man since early times are still important today and form the basis of further technical developments. Originally man was only interested in elements and minerals with particular properties. In the 19th century the sources of energy available from the caustobioliths (i.e. combustible raw materials of organic origin such as coal, crude oil, and gas) became particularly important. Since the end of World War II nuclear energy available from uranium, contained in some radioactive raw materials, has added to man's energy sources.

As a result of intensive scientific investigations, the consumption of mineral raw materials in modern industries has been increasing rapidly, both quantitatively (by worked volumes) and qualitatively (by the use of new minerals previously considered unimportant). Studies of the physics of solid substances, such as crystals, have led to new scientific discoveries which form the bases of further developments in the electronics, optics, and other industries. Specific properties of certain crystals gave rise to the origins of transistors, lasers, memory systems, efficient calculators, holography, and to a gradual miniaturization of modern plant instrumentation.

The importance of some minerals is indicated by the term strategic raw materials; these have often been the cause of international conflicts and wars.

Also the aesthetic properties of minerals led to their widespread use in human societies. Archaeological discoveries of various jewels ornamented with different precious stones provide good evidence that these aesthetic properties have always been of importance. Special attention has always been paid to minerals having extraordinary geometrical forms, bizarre patterns, splendent lustre, attractive colours, transparency and clarity, or an unusual play of colours, hardness and resistance against abrasion and damage. Because of these properties, brought out by cutting and polishing, and their rare occurrence in nature, these minerals are classed as precious stones and are highly prized. Varieties of inferior quality are used as ornamental stones.

This book is designed to provide the layman with a basic knowledge of mineral sciences, to describe the general character of some minerals and their varieties (130 mineral species are described and illustrated), to elucidate the rock-building processes, and to describe the best way to store and look after a mineral collection.

WHAT IS MINERALOGY?

Mineralogy is a branch of the natural sciences dealing with minerals. Its name is derived from the Greek *minera*, ore, and *lógos*, reason. Its beginnings as a separate science date from the 16th century. It is divided into two parts, general mineralogy and systematic or special mineralogy. *General mineralogy* is the science which defines minerals, describes their properties, and studies the mineral-building processes in nature. *Systematic mineralogy* is the science which establishes a mineralogical system based upon the chemical composition and internal structure of individual mineral species.

General mineralogy includes several separate scientific branches. *Descriptive mineralogy* is historically the oldest branch. It studies the variability of mineral species and whole accumulations, and the mode of their occurrence in nature. Acquired knowledge in this field has led to a special terminology which should become familiar to every collector of minerals and rocks.

At the end of the 19th century *morphological (geometrical) crystallography* became the most exact mineralogical discipline, and laid the basis for the development of structural and physical mineralogy. It studies the laws of the external geometrical arrangement of crystals. *Structural crystallography* studies the arrangement of particles within a mineral — the internal symmetry and structure of crystal lattices. *Crystallochemistry* is concerned with the relationships between the chemical composition and the structure of minerals, and *minerogenesis* (or *genetic mineralogy*) explains the principles governing the processes of mineral origin. *Mineral ontogenesis*, a new branch of genetic mineralogy, studies the origin of individual crystals and aggregates.

Experimental mineralogy studies the physico-chemical conditions of the origins of minerals, duplicating many of the natural processes in the laboratory. All the above-mentioned scientific branches use information from *geochemistry*, a science studying the laws of occurrence and migration of chemical elements on the Earth. *Physical mineralogy* is concerned with those properties of minerals that are due to the composition and structure of matter in the solid phase. In the course of the last 25 years this particular scientific branch has achieved the leading position among the mineralogical scientific branches. It is closely linked with mineralogy, petrology, geochemistry, and solid phase physics.

ELEMENTARY MINERALOGICAL TERMS

Minerals (with the exception of mercury) are chemically and physically homogeneous inorganic solid substances whose composition can be expressed by a chemical formula. In this way they differ from rocks (e.g. granite), which are usually a mixture of minerals. However, rocks may be formed of a single mineral which builds whole geological masses. Examples include limestone and quartzite. The study of rocks is called *petrology*. Minerals are natural substances and do not include artificial compounds (e.g. the materials formed on walls of blast furnaces), or synthetic crystals, or fossils (whose biological features are quite distinct). Minerals originated from magmas, solutions or gases, either in the form of *crystals* or *aggregates* (i.e. aggregations of irregular grains). Crystals may be *free* (not attached to a base), *embedded* (grown in a solid medium), or *attached* (growing from a solid base). They occur isolated or in groups.

Isolated crystals are called *monocrystals*. *Compound crystals* are formed by two or more individuals with a fixed orientation which may be in the form of *multiple twins* (gypsum), *contact* or *juxtaposition twins* (aragonite), *cyclic twins* (rutile), or *penetration* or *cruciform penetrating twins* (staurolite). A group of closely clustered crystals growing from a common base is called a *druse* (figure 1). Crystals lining an oval or globular cavity and pointing with their free ends to its centre are known as a *geode*. According to their shape (habit) we distinguish

Fig. 1. A druse of quartz crystals

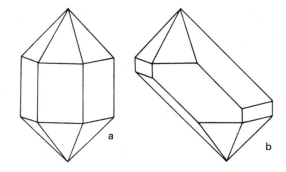

Fig. 2. Quartz crystals; combination of two rhombohedra and a prism; (a) of perfect development, (b) of one-directionally distorted development

isometric crystals, i.e. crystals developed equally in all directions (e.g. garnet); *tabular* or *lamellar* crystals, i.e. crystals developed predominantly in two directions (e.g. mica); and *linear, columnar* or *needle-shaped (acicular)* crystals developed solely in one direction (e.g. beryl, antimonite, and millerite).

Aggregates can be either finely or coarsely granular and *compact, stalky* (antimonite), *fibrous* (asbestos), *scaly* (haematite-specularite), *massive* (sphalerite), or *dendritic* or *flatly arborescent* (psilomelane). Apart from these compact aggregates porous, felt-like asbestiform, earthy or pulverulent forms may also be found. A piece of mineral or rock found in a natural locality is called a *specimen*. A *pattern* is a piece of mineral or rock collected for analysis.

WHAT DOES THE CRYSTAL SHAPE REVEAL?

Crystals are bodies bounded by usually flat faces arranged more or less symmetrically on a definite plan which is an expression of the internal arrangement (the *space lattice*) of their smallest particles — ions, atoms, and molecules. Crystals grow by apposition, i.e. the material is supplied to their core or older faces from the outside. In this way they differ from biological tissues which grow by interposition, i.e. by the intergrowth of younger cells between the older ones.

The regular internal arrangement of the crystal's space lattice is closely related to the external geometrical form of the crystal. A crystal, however, may grow freely only if it is isolated or embedded in a yielding or soft environment. Perfect and regular crystals (idiomorphic) are rare in nature; most often crystals occur in distorted and irregular forms (figure 2). But in spite of their irregular growth or different size, the *interfacial angle*, i.e. the angle between the identical two faces in all crystals of the same mineral, remains constant under the same conditions. This quality is defined by the *Law of Constancy*

9

of Angle, formulated in 1669 by the Danish naturalist Niels Stensen. Interfacial angles are measured by *goniometers*, either contact or reflecting, based upon the light reflectoin from crystal faces. Using these angles it is possible to determine the symmetry of the crystal and further specific features.

Crystals are three-dimensional bodies, and the mutual position of their faces is expressed by means of the triaxial coordinate system. With the zero point of this system in the centre of the crystal, the so-called *axial cross* results. The values of each axis, as well as being related to the crystal form, are also related to the maxima of certain physical properties of crystals in a given direction (hardness, electric and thermal conductivity, etc.).

From the morphological point of view, crystals may be divided into seven major divisions differing by the type of their axial cross. We call them *crystal systems*. They are the following: triclinic, monoclinic, orthorhombic, tetragonal, hexagonal, trigonal and cubic. Axial crosses of individual systems differ in the length of their axes and the angles of their intersections. Crystals of nearly all minerals are symmetrical to a certain extent; a complete absence of symmetry is rare.

The symmetry of a crystal is indicated by the symmetry elements, i.e. the centre of symmetry, the axis of symmetry, and the plane of symmetry. There are in all 32 crystal classes (including one asymmetrical class) into which all crystals of the various minerals (and also of artificial substances) may be arranged. Every crystal system has a characteristic common *basic form*, i.e. the form resulting from the connecting of the tops of its axial cross. Crystallographic systems with axial crosses and basic crystal forms are shown in figure 3. In the following description only the basic characteristic features of individual crystal systems are mentioned. More detailed data, such as the symmetry elements of individual classes, are beyond the scope of this book.

Triclinic system The axial cross is formed by three unequal axes (a, b, c) but none forms a right angle with any other. The system is divided into two crystallographic classes, one asymmetrical and one with the centre of symmetry. Minerals in this system include e.g. plagioclase, microcline, disthene, pectolite, amblygonite, and chalcanthite (blue vitriol).

Monoclinic system The axial cross is formed by three axes (a, b, c) of unequal length, two of which are at right angles. The third axis is inclined to the vertical plane. The system is divided into three classes, of which the class with the highest symmetry has the centre, the plane and the axis of symmetry. This system includes e.g. orthoclase, micas, augite, amphibole, titanite, and wolframite.

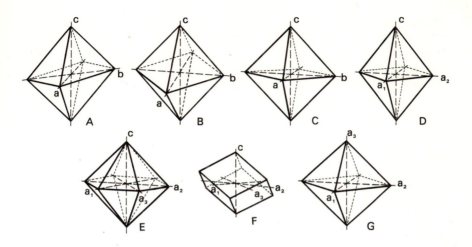

Fig. 3. Axial crosses and basic crystal forms of the crystal systems: (a) triclinic, (b) monoclinic, (c) orthorhombic, (d) tetragonal, (e) hexagonal, (f) trigonal, (g) cubic

Orthorhombic system   The axial cross is formed by three axes (a, b, c), all at right angles but of different lengths. It is divided into three classes, of which the one with the highest symmetry has the centre, three planes and three axes of symmetry at right angles. This system includes among others the following minerals: barytes, aragonite, cerussite, arsenopyrite, marcasite, antimonite, hemimorphite, olivine, and topaz.

Tetragonal system   The axial cross is formed by three axes at right angles (a_1, a_2, c), two of which (the horizontal a_1 and a_2) are of equal length. It includes seven classes; the class with the highest symmetry has the centre, five planes and five axes of symmetry. Minerals crystallizing in this system include cassiterite, rutile, scheelite, chalcopyrite, and vesuvianite.

Hexagonal system   The axial cross is formed by four axes. Three (a_1, a_2, and a_3) are of equal length and at 60° to one another, the vertical (c) axis is at right angles to the others. The system is divided into five classes; the class with the highest symmetry has the centre, seven planes and seven axes of symmetry. The following minerals crystallize in this system: beryl, apatite, 'high' quartz (beta quartz), nepheline, pyrrhotite, molybdenite, niccolite, wurtzite, etc.

Trigonal system (rhombohedral)   The axial cross is the same as in the hexagonal system, but the vertical axis has only a triple symmetry. It is divided into seven classes; the class with the highest symmetry has the centre, four planes and four axes of symmetry. Many common minerals crystallize in this system, e.g. calcite, dolomite, siderite, 'low' quartz (alpha quartz), corundum, tourmaline, haematite, cinnabar, proustite, pyrargyrite, graphite, and native elements such as arsenic and bismuth.

Cubic system The axial cross is formed by three axes of equal length and at right angles to one another (a_1, a_2, a_3). It is divided into five classes; the class with the highest symmetry has the centre, nine planes and thirteen axes of symmetry. This system includes the following minerals: halite, fluorite, garnets, magnetite, pyrite, galena, sphalerite, tetrahedrite, cuprite, and some native elements such as gold, silver, copper, and diamond.

Crystals bounded by faces of a single crystal form (e.g. hexahedron, octahedron or tetrahedron) are called *simple forms*. Crystals limited by two or more kinds of simple forms are called *combinations* (figure 4). Some crystal forms are typical of a particular mineral, e.g. an octahedron is typical of magnetite, an orthorhombic dodecahedron of garnet, and a hexahedron of halite. Other minerals may crystallize in many different forms. Calcite, for instance, crystallizes in 1,200 different forms, basic as well as derived. The crystal form is an important distinguishing feature. According to the form of crystals it is possible to distinguish, for instance, amphibole from the very similar pyroxene, magnetite from haematite, tetrahedrite from bournonite, rock crystal from clear gypsum or aragonite, vesuvianite from garnet, and many more minerals from others of similar appearance.

Certain crystalline substances are able to form several separate minerals, which are known as crystalline modifications. Crystalline modifications of the same substance have an identical chemical composition but different physical properties. They crystallize in different

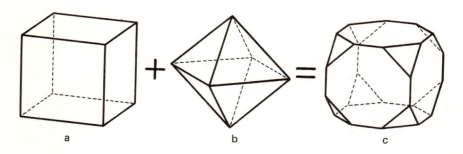

Fig. 4. Origin of combination (c) of two simple forms — cube (a) and octahedron (b)

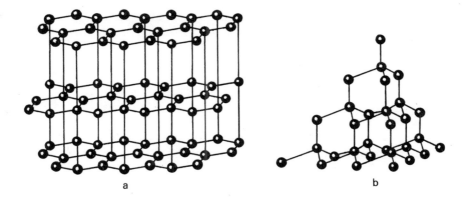

Fig. 5. The space lattices of two modifications of native carbon: (a) graphite, (b) diamond

crystal systems, and thus display different crystal forms; the crystal form is the best distinguishing clue (figure 5). For instance, calcite crystallizes in the trigonal system, aragonite in the orthorhombic; however, chemically both are calcium carbonates. Pyrite is cubic, marcasite orthorhombic, but both are bisulphides of iron. Another example of two chemically identical minerals differing in appearance and physical properties are the two modifications of pure carbon: diamond (cubic) and graphite (hexagonal).

A peculiar class of false crystal forms are the *pseudomorphs*. They originate secondarily, i.e. the original particles of the mineral were replaced by new particles, while the crystal form of the original mineral has been preserved. They may be the result of decomposition of an older mineral, and its alteration into a secondary product, or of hydrothermal replacement of a more soluble original mineral by another less soluble mineral. In case of an incomplete alteration, the relics of the older mineral form the core of the pseudomorph. Sometimes it is difficult to distinguish the pseudomorph from a mere coating by a younger mineral from external appearances only.

There are many different kinds of pseudomorphs in nature; examples include the pseudomorphs of limonite after pyrite or siderite, of malachite after native copper, of cuprite after azurite, of quartz after barytes, and of pyrite after calcite.

If the chemical composition of a mineral remains unchanged in the process of its alteration into another mineral, it is called *paramorphosis*. Examples include aragonite into calcite, marcasite into pyrite,

13

and occasionally diamond into graphite (never the other way round). Every paramorphosis is the result of a change in the structure of the space lattice (figure 5) of a given mineral.

MINERAL INTERIORS

Since early times, philosophers have thought that substances are composed of indivisible parts. This theory was further developed in the 18th and 19th centuries, by analytical, stoichiometric and organic chemistry, and by solid state crystallography. This hypothesis, however, was only substantiated as late as 1912 by the German physicist Max von Laue by a well-known experiment, in which a photographic plate showed patterns formed by X-rays passing through a monocrystal (figure 6). The experiment proved at the same time that X-rays are in fact an electromagnetic wave motion of ultrashort wavelength, and not a flow of individual particles (corpuscles). This significant discovery, which proved the atomic structure of substances, enabled a direct investigation of their microstructure and gave rise to a new science called structural crystallography.

As explained earlier, solid state substances may be divided into amorphous (of no regular shape) and crystalline (with a regular internal structure) forms. There is a considerable difference between the terms 'crystalline mineral', related to the internal structure with space lattice, and a 'crystallized mineral' (of typical external features). Every substance contains some internal energy and tends to decrease it as much as possible. From this point of view, the crystalline state is the most advantageous, and this is the common state of solids.

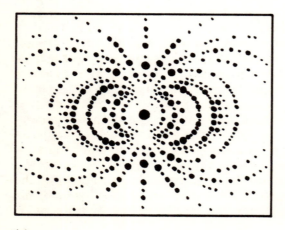

Fig. 6. Laue diffraction pattern — Roentgenogram. Photograph of diffraction of Roentgen rays transmitting a monocrystal.

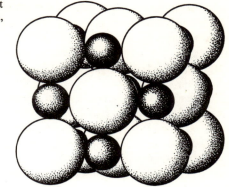

Fig. 7. Model of space lattice of rock salt (halite, NaCl): Na^+ — cations of sodium, Cl^- — anions of chlorine

Na⁺

Cl⁻

In amorphous minerals the arrangement of atoms, ions and molecules is quite accidental, irregular and chaotic. Modern science describes them as 'subcooled liquids'. Amorphous minerals are *isotropic*, i.e. they behave in the same manner in all directions. Isotropic minerals usually are *reniform* (kidney-like) or *globular*. Some opals, however, although typically amorphous minerals, have a regular and symmetrical internal structure. Precious opal is composed of large globular aggregations of amorphous silica gel (silicon dioxide with water) of a uniform diameter of tens of micrometres arranged into a regular skeleton. Their geometrical arrangement is the physical cause of the characteristic play of colour in precious opals.

Crystalline minerals are characterized by a regular internal arrangement of their space lattice (figure 7). The external expression of the internal arrangement is a crystal. Crystals, unlike amorphous substances, exhibit different properties in different directions, i.e. they are *anisotropic*. The best proof of anisotropy is the crystal shape. If a crystal displayed identical properties in all directions it would have the shape of a globe.

The symmetry of the internal arrangement of crystals and crystalline substances can be expressed by seven elements of symmetry. This symmetry may be deduced by means of a point chosen in the space and shifted along axial crosses of the crystal systems. As a result, 14 elementary translation lattices originate. By the combination of those lattices according to the seven elements of internal symmetry 230 space groups originate.

The type of elementary translation lattice in which a mineral crystallizes is determined by the relative size of atoms, ions or molecules, and their mutual chemical bonds. The internal arrangement of the elementary lattices determines the external form of the crystal. If the elementary lattices are interconnected in all directions, the crystal is

15

isometric (e.g. diamond); if flat interconnection prevails, the crystal is usually tabular or lamellar (e.g. graphite and mica); in linear interconnection the crystal forms columns or fibres (e.g. pyroxenes and asbestos). The internal arrangement of the elementary lattices also determines different physical properties of crystalline minerals, i.e. hardness, cohesion, cleavage, and fracture.

The smallest unit of a crystalline space lattice or the basic configuration of every crystalline substance is the so-called *basic parallelepiped.* The lengths of its edges marked with a_0, b_0, c_0 are known as the basic *lattice parameters* (lattice constants) of a mineral. They give the shortest distances between atoms (or ions or molecules) in three directions forming the basic parallelepiped, and are the most important specific features of every crystalline mineral.

If a crystalline substance grew regularly and had a perfect arrangement of its space lattice, it would achieve the form of an *ideal crystal.* But in nature as well as in common laboratory conditions only the so-called *real crystals* arise, characterized by various distortions, inclusions or interruptions in their structure. Only if the number of these is negligible does an almost *perfect crystal* result. A perfect external arrangement of crystal faces does not, therefore, prove a perfect internal structure.

A *crystal defect* includes all spot, linear, areal and volume irregularities affecting the chemical or geometrical homogeneity of the internal structure. It includes impurities, the absence of some particles in the space lattice (lattice vacancies), excess particles wedged in the space lattice, or different microdislocations causing the failure of the internal arrangement of a monocrystal into differently oriented blocks of mutual cohesion. A detailed investigation of all these crystal defects is of great theoretical as well as practical importance, since they cause many important physical properties of real crystals. A perfect or ideal crystal could never have such properties. Real crystals are used in semiconductors, diodes, and transistors.

The investigation of the internal arrangement of crystalline minerals is usually carried out by Roentgen structural analysis. It is based upon the diffraction, reflection and interference of X-rays transmitting details of the crystal lattice.

The mathematical equation established by two English physicists, the father and son Bragg, expresses the relationship between the wavelength of X-rays, the distance of structural planes in the space lattice, and the angle of the deflected X-ray. If two of these values are known, the third may easily be calculated. A systematic Roentgen structural analysis of all crystalline minerals has provided important data not only on their internal arrangement, but has also enabled the measurement of real atom and ion diameters and also measurement of the size of different molecules.

MINERAL PROPERTIES

The physical properties of minerals result from both the properties of atoms of a particular element and the properties of the whole crystalline lattice. The first group of properties includes radioactivity and also, to a certain extent, luminescence. The second group includes mechanical and electrical properties. Sometimes, however, the real cause of certain properties still remains unknown.

Minerals may exhibit *scalar properties*, such as specific gravity, density, and specific heat, and *vectorial properties*, such as hardness and magnetism. Some physical properties of certain minerals may also vary according to the direction from which they are handled. Consequently, we may say that crystals are anisotropic bodies. Every mineral has some typical properties which are constant and are an important aid in its identification and these types of properties are described below.

The **specific weight** is the weight of 1 cm³ of a mineral expressed in grams.

The **specific gravity** (or relative density) of a mineral is the relative number indicating how many times heavier or lighter a mineral is compared with the same amount of water at 4 °C. Due to different impurities and inclusions the specific gravity may vary considerably.

Hardness is the power of resistance to mechanical injuries. It may be indicated relatively, or for an exact determination of hardness a sclerometer is used. In some minerals (e.g. disthene) the same plane of the mineral may resist scratching in a different manner according to the direction—longitudinal or transverse—in which it is scratched. For the determination of hardness, a hardness scale was established, called *Mohs' scale of hardness*. It has ten grades in which each higher grade of mineral can scratch the one just below it in hardness. The set of standard minerals used as examples for the hardness test is as follows: 1 — talc; 2 — rock salt (halite), gypsum; 3 — calcite; 4— fluorite; 5 — apatite; 6 — feldspar (orthoclase); 7 — quartz; 8 — topaz; 9 — corundum; 10 — diamond. The scale is only useful for comparison, the figures do not indicate how many times harder a given specimen is than some other member of the scale. The *tenacity* of a mineral is its ability to resist pressure, tension, or torsion. If a mineral is subjected to pressure, etc., beyond its cohesion, it may suffer permanent clastic deformations (*fracture* and *cleavage*) or plastic deformations (*flexure* and *gliding*).

Optical properties are the reaction of a mineral to light. With the naked eye we may notice that the light rays are either reflected from the mineral planes — the mineral is lustrous; or they are selectively absorbed — a certain definite colour results; or they travel through the mineral — it is transparent or clear. Only some minerals have their own constant colour, unchangeable even under different conditions, e.g. malachite and azurite. Many minerals are varicoloured, e.g. fluorite and tourmaline. Some minerals display different colours when viewed in different directions by transmitted light. This property is called *pleochroism*. Cordierite, for instance, displays three different colours in three different directions at right angles to one another.

In the identification of minerals the *streak* may be of much importance, being constant irrespective of colour variations. The streak is the colour of a fine powder left behind after the mineral has been rubbed on a piece of roughened or unglazed porcelain.

A ray of light is refracted when transmitted through a mineral. The mathematical relationship between the angle of the light incidence and the light refraction is called the *index of refraction*, and is an important and specific feature of every mineral. *Light polarization* is a complicated physical phenomenon which may be studied only by means of special apparatus such as a polarizing microscope. It is used especially in petrology in the study of thin sections of rocks.

Magnetism is the property of minerals which contain in their space lattice atoms with the so-called magnetic moment, with unpaired electrons on external orbits. The mutual orientation of these atoms establishes the magnetic structure of the crystal lattice composed of magnetic domains, and hence of the whole crystal or aggregate. *Diamagnetic* minerals are non-magnetic and cannot be magnetized (e.g. copper); *paramagnetic* minerals can be magnetized only slightly (e.g. pyrite); and *ferromagnetic* minerals are always magnetic (e.g. magnetite and pyrrhotine). In nature some minerals may be magnetized by the effect of a strong magnetic field. Rocks containing magnetic minerals may become magnetic, for instance, after a lightning stroke. Magnetic properties of minerals are of great importance. Field magnetometry is used in geophysical prospecting for iron ore deposits. Magnetic separation is used in ore dressing, the magnetism facilitating the determination of minerals, e.g. distinguishing magnetite from the similar chromite. Magnetic minerals are used in electronics and many other industries.

Palaeomagnetism is of special importance in geology. It is based on the relationship between the orientation of the magnetic axis of a crystalline mineral as it is formed and the magnetic field of the

Earth. To study palaeomagnetism, the position of a rock specimen must be exactly determined by means of cartographic coordinates. At the time of crystallization of the specimen its magnetic axis always points to the north pole. Thus, by using accurately positioned specimens and samples dating from different geological epochs, the actual position of magnetic poles, their migration in the course of the historical evolution of the Earth, or the migration of whole continents may be determined. Conversely, from the orientation of the magnetic axis the geological age of the specimen may be judged.

Electrical properties become evident when the mineral is placed between two poles of electric voltage. These properties show the presence of free electrons and ions and depend upon the possibility of the migration of these electrons and ions within the space lattice. The decisive factor of this migration is the type of chemical bond, i.e. the manner in which the particles are bonded. According to the mobility of electrons within the lattice, minerals are divided into three groups: insulators — which are non-conducting, conductors — which contain free electrons and are conducting, and semiconductors — which are insulators conducting electric charge only in certain circumstances. The investigation of semiconductors resulted in discoveries which became important in the production of transistors, microrectifiers, computers and microprocessors. Semiconductors led to miniaturization and to the automation of control of complex processes.

If a mineral has a polar crystallographic axis — displaying different properties on each end — it may become the source of static charge and exhibit potential difference. This occurs when the electrically neutral elementary structural unit in the lattice is distorted by some external interference, e.g. mechanically or by heat, and consequently turns into a dipole. *Piezoelectricity* originates by the deformation of the space lattice due to pressure. The space lattice becomes elongated or shortened in the direction of the polar axis, and the ends of the axis become electrically charged. On the other hand, if the ends of the polar axis are charged by alternating current, the crystal becomes alternately elongated or shortened. An example of a piezoelectric mineral is a crystal of quartz. Quartz plates are used in apparatus to obtain ultraoscillations. *Pyroelectricity* originates in some crystals with a polar axis (usually vertical) as a result of heating. When heated, they develop opposite charges at opposite ends of the crystal, and attract, for instance, the ash from a fire. Examples include tourmaline and hemimorphite.

The **thermal properties** depend upon the heat conductivity of the mineral, which is variable in different directions, as is thermal expansivity. The thermal energy of the lattice is caused by oscillations of its particles. If the amplitude of these oscillations exceeds the distance of the interacting bonding forces, the mineral melts or volatilizes.

Radioactivity is only found in minerals which contain radioactive elements of the decay series of uranium, thorium, actinium, or trace amounts of radioactive isotopes, such as potassium. The radioactive decay causes changes in the chemical composition of the mineral, and the radiation destroys the crystal lattice. Some minerals reveal pleochroic halos, e.g. biotite around inclusions of zircon. Other minerals change into metamictal minerals, such as betafite and samarskite. Their original crystal lattice becomes amorphous, yet their external crystal form remains unaltered. From the quantity of the final products of the radioactive decay of elements and from the disintegration rates (called half-life periods) the age of minerals may be calculated. In this way, also, the geochronology of the historical evolution of the Earth may be determined. Radiometry is a highly efficient geophysical method of searching for deposits of radioactive raw materials.

Luminescence, a property of certain minerals, is caused by the change into light rays of various kinds of energy (e.g. thermal, luminous, electrical, chemical, and mechanical). In mineralogy the most important is *photoluminescence* originating after some minerals have been exposed to ultraviolet light of a wavelength of 253.7 or 336 nanometres. It occurs only when the mineral contains a trace amount of an activator, usually manganese, copper, lead, silver, uranium, elements of rare earths, water molecules, etc. Since only some specimens of the same mineral may usually produce coloured luminescence, it cannot be used as a reliable clue in mineral identification.

RECENT ADVANCES IN OUR KNOWLEDGE OF MINERAL PROPERTIES

Many properties of minerals may be explained by the change in the behaviour of free atoms after they enter the solid state. These are the properties resulting from the electron structure of atoms. Apart from chemical properties, they include the mutual bonding of atoms, ions, and molecules in the mineral, and some of its physical properties. Atoms in the crystal lattice influence and adapt to one another. Characteristic features of a mineral are given by its bond and space lattice. It may be deduced, therefore, that mechanical, optical, electrical and magnetic properties of minerals are the properties of electrons in compounds.

The modern physical crystallography is based upon the theory of chemical bonds. There are three main types of chemical bonds: *ionic bond* (heteropolar), *covalent bond* (homeopolar) and *metallic bond.* Related to these, three theories, called the crystal field theory, the molecular orbital theory, and the band theory were formulated.

The *crystal field* is electrostatic in nature and occurs in minerals with ionic bonds, in which some properties and the behaviour of every ion and its electrons are determined by the number and position of the nearest ions (ligands).

The molecular orbital theory does not study the migration of electrons on orbits of one atom, but on orbits common to two or more atoms. In minerals it may be applied in complex building particles with a covalent bond of atoms (ions) in a molecule, e.g. in anions (carbonate, sulphate anions, etc.), in sulphur molecules in the space lattice, etc.

The band theory is concerned with the migration of electrons inside the space lattice. It distinguishes energy levels in a crystal analogous to electron orbits in the atom, i.e. the valence band and the band of conductivity. If they overlap, the mineral is a conductor. If they are divided by the so-called band gap (zone without orbits), the mineral is a semiconductor. If the gap is filled with an element (the 'acceptor' or 'donator' of electrons), the mineral is a doped semiconductor. In case of a wide band gap the mineral is an insulator.

The theories outlined above may explain many different physical phenomena including the *colour of minerals.* Colour is caused by a selective absorption of some wavelengths of transmitted light. According to the degree of absorption, minerals are divided into four groups. Colourless, clear (achromatic) and transparent minerals transmit light without any selective absorption; an example is rock crystal. Idiochromatic (coloured) minerals contain a certain proportion of a colour-forming element called a chromophore, e.g. copper in azurite. Allochromatic (coloured) minerals contain the chromophore as an admixture only in trace amounts, e.g. chromium in rubies and emeralds. A special group is formed of minerals coloured by very small foreign mineral particles (microinclusions), e.g. chlorite and haematite in jasper. Seemingly coloured (pseudochromatic) minerals display colour effects which are due to refraction, reflection, or interference of light, e.g. the 'fire' of the diamond and the opalescence of precious opal. However, this classification does not elucidate the fundamental causes of mineral colouring.

For instance, copper was believed always to produce green or blue, chromium red, cobalt deep blue. In fact compounds of chromium may be red, orange, yellow, green, or violet. A deep blue colour of almost identical tinge may be caused by iron and titanium in sapphires, by cobalt in spinels, by copper in shattuckite, by sulphur in lazurite, and the so-called 'colour centre' due to irradiation may be the cause of the blue colour in certain beryls. These phenomena are studied in physical crystallography by a complex investigation of every specimen, and by a subsequent application of one of the three previously mentioned theories of electron migration in solid substances.

The crystal field theory elucidates the origin of colours in minerals in several ways. According to one of them, transition metals, which are found in the minerals listed, cause their characteristic colours.

Chromium: green—uvarovite and grossularite (garnets), emerald, hiddenite, verdelite; orange—crocoite; red—ruby, spinel; red with green—alexandrite (chrysoberyl).

Manganese: pink—rhodochrosite, rhodonite, rubellite, morganite, kunzite; orange— spessartine; yellow—ganophyllite.

Iron: blue—lazulite, indigolite; green—olivine, verdelite, demantoid, andradite; red—pyrope, almandine, ludlockite; yellow—goethite, citrine.

Cobalt: pink—erythrite, rosalite, spherocobaltite.

Nickel: green—chrysoprase, bunsenite.

Copper: blue—azurite, chrysocolla, turquoise; green—malachite, dioptase; red—cuprite.

The colour of a mineral may be also caused by the colour centres or defects in the space lattice, usually due to the irradiation of the mineral by gamma rays. Minerals having colour centres include fluorite, smoky quartz and morion, irradiated diamonds (of various colours), and brown topaz, and probably also blue and yellow halite, blue barytes and coelestine, and yellow calcite.

According to the molecular orbital theory, the colour of a mineral is due to the transmission of electrons between ions of two identical or different elements, particularly metals. For instance, blue sapphire and benitoite have electron transmission between iron and titanium, vivianite has electron transmission between ions of iron, covelline has electron transmission between ions of copper, and pyrite, marcasite, skutterudite, sylvanite, and others have colour caused by electron transmission between metals and non-metals.

According to the band theory, the colour of minerals — conductors or semiconductors — is caused by the selective absorption of light by electrons migrating between the valence and conductivity bands, or in the band gap. In mineral insulators the colour may be due to the admixture of a particular element, filling with its orbits the band gap in the mineral, and changing the insulator into conductor. For instance, an admixture of boron colours the diamond blue and makes it conductive (an example is the famous diamond called 'Hope').

Thus, both modern physical mineralogy and crystallography have contributed greatly to the elucidation of the causes of many of the physical properties of minerals, and have led to practical applications.

MINERAL ORIGINS

The minerals and rocks of which the Earth is formed originated in all stages of its historical development. The science studying the origin and occurrence of minerals is called minerogenesis. Minerals originated in the course of many different processes, sometimes one mineral may have originated in several ways, and only rarely can the origin of a mineral be directly traced. Important clues to mineral origins are acquired from studies of active volcanoes or thermal springs, or from investigation of minerals deposited on the bottoms of lakes and oceans, or in depleted raw material deposits. Laboratory experiments, the production of synthetic crystals, or accidental formation of crystals in industrial processes (metallurgical or chemical) are also of great importance. All this knowledge may be applied in the elucidation of different mineral-building processes in the Earth's crust as well as in the study of the mineralogical evolution of the Earth and in defining the laws governing the origin of minerals in nature.

Minerals are divided according to their origin into *primary minerals,* formed from molten magmas, hydrothermal solutions and gases, and *secondary minerals,* originating by the decomposition or weathering of primary minerals. A more exact division, however, would be into endogenous minerals originating from within the Earth — *hypogenic, ascendant minerals* — and *supergenetic, descendant minerals,* which are the result of the activity of surface (exogenous) agents, such as the atmosphere, precipitation, etc.

As a result of various mineral-forming processes, large accumulations of some minerals may produce *workable deposits of ore minerals.* According to their origin and form we distinguish vein deposits, i.e. cracks in the Earth's crust filled by younger ore-bearing and other minerals (gangue material), and stratified deposits, in which raw ores form layers, intercalations and impregnations concordant with the stratification of the rocks.

Some minerals are original parts of the rocks in which they are included (*syngenetic minerals*), e.g. olivine in basalt, and the layers of sedimentary pyrite in schists. Other minerals were deposited into older rocks from solutions as the outcome of younger mineralization (*epigenetic minerals*). For instance, most ore veins originated in cracks and fissures in the Earth's crust filled with minerals separated out from circulating hot solutions.

A separate group is formed by minerals which originated as a result of metamorphic processes. They are formed from older minerals which, under changed pressures and temperatures, lost their original stability and changed into some other minerals.

Another type of mineral alteration is the weathering of minerals. In the process of weathering certain elements are liberated from older minerals or rocks. Large accumulations of various raw materials are often formed, e.g. kaolin or bauxite deposits formed from feldspars and the so-called oxidation-cementation zones of ore veins ('iron hat'). Iron hats or gossans may be found in the weathering zone at the contact of ore veins with the surface of the Earth. At the level of the underground water, another zone, that of precipitation (cementation), may be found, produced by constituents dissolved in solutions circulating near the surface. Cementation zones are usually enriched with gold, silver, and some other metals, whose content in unweathered and deep-seated parts of veins is very low and therefore unworkable.

Mineralization (mineral-forming) processes may generally be divided into a series of types. In one of the processes, *magmatic* mineralization, minerals are formed from magmas cooling far underground. The hot molten magma, originally containing large amounts of almost all chemical elements, differentiates gradually into partial magmas of different compositions. As the temperature or pressure decreases various igneous rocks separate out. The remainder, called the *residual magma*, contains water vapour, volatile gases (such as fluorine and chlorine), and concentrations of certain elements (which were much more dispersed in the original magma). The residual magma gives rise to hydrothermal solutions seeping into rocks situated nearby. According to their solubility, minerals are precipitated from solutions circulating through fissures and cracks. They form hydrothermal veins or impregnate porous rocks. By the replacement of soluble rocks, such as limestones, with ores and mineral raw materials, they formed extensive deposits of *metasomatic origin.*

Sedimentary mineralization gives rise to chemical sediments, e.g. deposits of rock salt, gypsum, and borates. Various micro-organisms usually participated in their origin.

Metamorphic mineralization includes two main types of mineral-forming processes. In one the molten, fluid magma often penetrates and invades the surrounding rocks which are affected by heat, especially at the contact with the magma. Minerals originating along the contact line are termed contact minerals; the process is called contact metamorphism. A second type of metamorphic mineralization involves large rock masses, originally formed on the surface of the Earth, then buried to great depths, where they are subjected to great pressure and high temperatures and are altered on a regional scale to metamorphic rocks. This is the way in which mica schists or gneisses originate. Metamorphic processes also include the transformation of certain igneous rocks (e.g. dunites and dark volcanites) into serpentine. The rock volume increases in this process.

The origin of minerals is by no means accidental but is governed by certain definite laws. Therefore the *occurrence of minerals* (the locality of occurrence) is related to the geological environment and its development. In mineralization processes sometimes only one common mineral is produced, such as limestone, gypsum, quartz, or rock salt. More often several minerals appear associated and form the characteristic association called a *paragenesis*. Typical examples are igneous rocks such as granites, composed of quartz, feldspar, and mica. Associations of minerals occurring in ore veins were known to miners of old.

Metalogenesis is a branch of science involving the study of the origin of ore deposits. It distinguishes several parageneses important from the viewpoint of raw materials. They are as follows: gold-bearing formations composed of native gold, arsenopyrite, and pyrite in quartz veins; quartz veins containing antimonite and native gold; tin-wolfram-molybdenum ores in quartz veins with topaz, zinnwaldite, fluorite, apatite, and other minerals; formations of polymetallic ores of silver-lead-zinc in quartz-carbonate veins; the formation of nickel-cobalt-bismuth-silver-uranium in carbonate veins; the siderite formation with copper and mercury ores, and some others.

The 'Alpine paragenesis' is a typical association of minerals filling cracks in some crystalline schists. These rocks were the sources of elements extracted by circulating solutions under high temperatures and great pressures, which, if in sufficient quantities, precipitated as minerals forming rich druses, e.g. of rock crystal, albite, rutile, and haematite.

The mineral composition of all parageneses is predetermined by geochemical factors. Each paragenesis may either contain just its typical minerals, or also minerals occurring in other different parageneses. Minerals originating in different mineralization processes, however, can never crystallize associated. For example, garnet never precipitates with gypsum, nor galena with diamond, nor feldspars with rock salt. In a particular locality all minerals of the same age and the same mineralization process belong to one *generation*, in which the *successive occurrence* of minerals may be distinguished. In one deposit a mineral may occur in several generations dating from different mineralization phases. The knowledge of parageneses as specific mineral associations is of great practical importance.

STORAGE PROBLEMS DUE TO THE INSTABILITY OF MINERALS

In nature as well as in collections, minerals, as unstable natural materials, are subject to different changes and sometimes may even be destroyed. If the characteristic features of a mineral are to be preserved, expert attention and care should be paid to the collection. The surfaces of many minerals may be affected physically as well as chemically. Physical agents include light, alternating periods of heat and cold, and humidity (which favours chemical decomposition). One of the most significant problems is the *change of colour*. The change of colour of a mineral is mostly caused by physical or physico-chemical agents. When exposed to the light or radiation, photosensitive minerals such as some amethysts, smoky quartz, brown topaz, fluorite, zircon, diamond, and hackmanite (sodalite variety) lose their colour. The chromophore in these minerals is formed by the colour centre. Their original colour may be regained again under certain special conditions, e.g. by heating or by radiating with gamma rays. There are, however, some silver-containing minerals which lose their colour for ever, e.g. proustite and pyrargyrite. Originally they are transparent and of a bright cherry-red colour. However, when exposed to the light they turn black and their adamantine (diamond-like) lustre changes to metallic. Probably these changes are caused by the photochemical effect under which silver is reduced in the space lattice of both minerals. Realgar also changes its appearance in the light, but this phenomenon has not, as yet, been explained. Its bright red colour changes through orange to yellow, it loses its adamantine lustre, and its surface becomes coated with a powdery mixture containing orpiment. The best way to protect mineral specimens from the harmful effects of light is to wrap them in non-transparent bags or to keep them in the dark.

A seeming change of colour is a chemical phenomenon. It is usually caused by a thin film of secondary products of weathering, accompanied sometimes by the loss of chemically bound water, or by its absorption from the surrounding air. Examples include ankerite which is normally white but which is coloured brown by limonite when decomposed; the whitish dehydrated encrustations on chalcanthite or melanterite; the tarnish colours of some sulphides; and the dull and softened surface of rock salt. To protect minerals from decomposition it is necessary to preserve them in a dry and inert environment which excludes the harmful chemical effects. The dehydration of a mineral will best be prevented when it is placed in a hermetically sealed transparent bag with a small inserted tuft of wet cotton wool; the same kind of bag (without the damp cotton wool) may be used to protect, on the other hand, various hygroscopic salts from humidity.

Chemical decomposition of minerals is very common. The most harmful for a collection of minerals is the decomposition of 'kiess', especially marcasite, pyrite, arsenopyrite, argentopyrite, sternbergite, and others. This decomposition is caused by the oxidation of sulphides in the presence of water. It takes place in several phases, and besides iron sulphates, pure sulphur and limonite, it also produces sulphuric acid. This dissolves other chemically less resistant minerals and gives rise to many secondary sulphates, e.g. gypsum, epsomite, and halotrichite. In the advanced stage of decomposition the mineral specimen crumbles (together with its wrapping, labels, and storage box); sometimes even the neighbouring specimens may be affected. This holds true even of specimens in which pyrites occur only in very small amounts.

The only efficient protection against 'kiess' decomposition is to conserve the specimens and keep them in a dry place. Specimens containing 'kiess' must be conserved before they are included in the collection, even if they seem perfect. Specimens once affected by decomposition may be saved only by a timely measure. First of all, weathering products must be mechanically cleaned from such a specimen with a brush. The specimen must never be washed with water. The remains which cannot be removed with a brush will dissolve in methylalcohol (caution: poison!) which, at the same time, will absorb the sulphuric acid. The operation should be carried out in a vacuum desiccator to enable the solvent to penetrate even the microcracks. The process should be repeated in fresh fluid until the methylalcohol, turning brown, remains clear. The cleaned specimen should be left to dry and then immersed in a conservation agent, e.g. some polymer of suitable quality and viscosity. Only inert polymers which remain clear as solids should be chosen. It is not advisable to varnish the specimen or to wax it.

The impregnation carried out again in the vacuum desiccator is complete when no more bubbles of escaping air rise from the bath. The final phase of the conservation, after the specimen is left to drip, can be shortened by placing it for 24 hours in an electric drier with a temperature not higher than 40 °C. The specimen becomes coated with a thin film of polymer which protects it from the chemical effects of the atmosphere, and reinforces its interior.

Minerals are sometimes damaged by other substances from the atmosphere, especially by carbon dioxide, sulphur and nitrogen oxides, dust, vibrations, and other harmful agents. Mineral specimens should therefore be kept in a safe place free from all harmful influences, if they are to continue to exhibit the brilliance and beauty of the mineral kingdom to the full.

SYSTEMATIC MINERALOGY

The mineral kingdom embraces about 3,000 mineral species, most of which have several varieties. A mineral species is a mineral of a definite chemical composition and a definite space lattice. Substances of an identical chemical composition but a different internal structure are therefore considered separate mineral species, e.g. graphite and diamond, calcite and aragonite, pyrite and marcasite.

The mineralogical system based upon the chemistry and structure of minerals associates species in families according to these features, and therefore it may be considered a natural system. The classification of a mineral is never definitive, however, since further investigation of its structure can result in additional changes in the original classification.

This book deals with 130 mineral species and their varieties, divided into the following classes of the mineral system (the page numbers are given in brackets): elements (30—45); sulphides and analogous compounds (46—87); halides (88—91); oxides (92—127); carbonates (128—145); sulphates (and chromates, wolframates, and molybdates) (146—155); phosphates (156—169); silicates (170—217); and organic compounds (218—221).

COLOUR ILLUSTRATIONS

Key to abbreviations used

H.   - hardness
Sp.gr.   - specific gravity
Melt.point - melting point

Copper Cu
Cubic; H. 2.5—3; Sp.gr. 8.5.—9; Melt.point about 1,350 °C

Copper is the most common native metal occurring in nature. The Earth's crust contains approximately 0.01 per cent copper. Apart from gold, it was one of the first metals used by primitive man, as early as the Late Stone Age. The name *cuprum* was given to it by the Ancient Romans, after its deposits on the island of Cyprus. Usually copper occurs as a component of primary or secondary ore-bearing minerals from which it is released by the process of reduction in its native state in the upper part of copper lodes found all over the world. The most important localities for the occurrence of copper are Rheinbreitbach, FRG; Chessy, France; Río Tinto, Spain; Moldova, Romania; Recsk, Hungary; Zlaté Hory, Czechoslovakia; the Urals and Altai, USSR; Ajo and Bisbee, Arizona, USA; localities in Bolivia and Chile; the African 'Copper Belt' in Zaire and Zambia; and in Bura-Bura Mine, Australia. Sedimentary copper is found in schists near Mansfeld, GDR, and in large deposits north-west of Wroclaw,-Poland.

Primary native copper originates in large quantities by crystallization from hot solutions together with melaphyre volcanic rocks, and in calcite veins in the locality of Keewenaw Point on the southern shore of Lake Superior, Michigan, USA. Of a similar origin is copper found in Långban, Sweden, and Franklin Furnace, New Jersey, USA.

Copper is used extensively in metallurgy (in the manufacture of bronze and brass), in electronics, in the chemical industry and in medicine. In the past it was used in the manufacture of cauldrons and for roofing.

1

Copper usually occurs in cube-shaped, octahedral or rhombo-dodecahedral crystals (2, 3) forming clusters, often one-sided in their growth, or skeletal (1, 4), dendritic, feather-shaped, arborescent or mossy aggregates. It very often occurs in the form of plates (5), tabular encrustations, grains, clumps and compact masses. Exceptionally copper blocks of up to 1600 kg have been found, e.g. in Vozniesensk, USSR (secondary copper formed in the cementation zone), and blocks of 1,500 tonnes have been found near Lake Superior (primary copper).

The typical bright copper-red coloration becomes dark on exposure to the light, the high lustre turns dull, and the surface becomes coated with secondary minerals. It has hackly (jagged) fracture, a metallic copper-red streak, is very ductile and malleable, and as a conductor of heat and electricity almost equals silver.

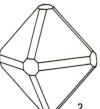

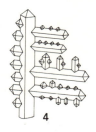

31

Silver Ag

Cubic; H. 2.5—3; Sp.gr. 9.6—12; Melt.point about 1,000 °C

Silver has occupied a special position in the history of mankind. Whereas gold became a symbol of riches and power, silver — above all a coinage metal — served for the acquisition of riches and power. Silver, mined since the 6th century B.C. in the Lavrion deposit, was the basis of the economic prosperity of Athens and all of Greece. Equally important for the Romans was the mining of silver in Hispania (present-day Spain).

The Earth's crust contains about 0.0001 per cent silver, but native silver forms only a small part of its total occurrence. Mostly it occurs in compounds together with sulphur, antimony or arsenic, or is microscopically dispersed in sulphides, e.g. galena. Native silver results from their decomposition, and as a secondary metal, it forms large accumulations with argentite-acanthite in the cementation zone of ore veins. Abundant deposits of this type may be found in Freiberg and Schneeberg, GDR; Wittichen, FRG: Kutná Hora, Jáchymov and Banská Štiavnica, Czechoslovakia; Cornwall, Great Britain; Chalanches, France; Blagodat, Urals, USSR; Broken Hill, Australia, and elsewhere. Rich deposits have been found in Zacatecas, Mexico; Chañarcillo, Chile; and Potosí, Bolivia. Primary native silver occurs in Kongsberg, Norway, and Příbram, Czechoslovakia; it is found as a component of copper in Lake Superior, Michigan, USA. Silver in the free state (disseminated) is found in copper-bearing schists near Wroclaw, Poland, and near Mansfeld, GDR, where the well-known zoomorphoses of fish were found (fossils replaced by silver). In the Middle Ages rich silver deposits in Germany and Bohemia were responsible for two-thirds of world production. At present some 80 per cent of world production comes from the two American continents.

3

Silver is found in the form of various many-faced crystals (3, 4), platelets, wires (1,2), skeletal aggregates, or dendritic, and compact nodules of silver-white colour with a yellow-brown tint. It is glossy, but becomes black and coated with argentite when exposed to the air. It has a hackly fracture, is flexible, ductile and malleable (a wire 1 km long can be drawn from 0.5 g of silver), and it is a very good conductor of electricity.

2

4

The world consumption of silver has been increasing. One-third of the total production is used in coinage, and a considerable amount of silver is needed for the production of silver jewellery. Consumption of silver has increased with the development of modern industries, such as the chemical industry, photography, metallurgy (special alloys) and electronics.

1

Gold Au
Cubic; H. 2.5—3; Sp.gr. 15.5—19.3

Gold was the first metal known to ancient man, and has been known since the Neolithic period, when it was mainly obtained from deposits of alluvial type. Ancient Egyptians mined gold as early as 2900 B.C. It was mainly used for religious purposes, and in the production of jewellery and decorative objects; later it was also used in coinage. It has preserved its dominant position in finance. The most ancient decorative objects made of gold known at present were found in 1972 in Grupina, near Varna, Bulgaria, and date from about 4000 B.C.

The Earth's crust contains a small amount of gold (about 0.000 002 per cent) mostly dispersed in various rocks (about 0.005 g/tonne). It becomes concentrated in mineral deposits and, after their weathering, accumulates as secondary deposits in alluvial material. Primary deposits originate as a result of magmatic activity, gold being separated out on hydrothermal quartz and carbonate veins, either in larger sheets or dispersed in associated minerals, such as pyrite, arsenopyrite, antimonite, etc. Exceptionally, it occurs in combination with tellurium and other elements, forming a special group of rare minerals. After the decomposition of the matrix gold sometimes dissolves and, under favourable conditions, is separated out in the enriched cementation zone of deposits. Tiny platelets of gold in geologically old alluvial deposits became part of sedimentary rocks in which they form gold-bearing layers of comparatively large extent, such as alluvial deposits at Witwatersrand, South Africa. Apart from being stored in state gold reserves, gold is used in jewellery (about 60 per cent), in the electrical industry (28 per cent) and in dentistry (10 per cent).

2

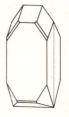

3

Gold usually occurs in octahedral, variously elongated crystals (1, 3). They usually grow into skeletal (4) or dendritic bunches or form sheets (2), often with the characteristic triangular striation (5), wires, or flakes. In alluvial deposits massive gold may be found. The largest nugget (153 kg) was found in 1851 in Chile. Gold is the most easily malleable metal, and an excellent conductor of electricity. Its typical gold-yellow colour changes with the admixture of silver and copper. The most important deposits are in South Africa (about 70 per cent of world production), with smaller deposits in the USA (Mother Lode, California, and Cripple Creek, Colorado), in Canada and the USSR (some 20 per cent of world production), and in Australia (Kalgoorlie, Coolgardie). In Europe, the most important deposits are in Romania (Roşia Montana, Brad, Baia de Arieş, and Zlatna).

Arsenic As
Trigonal; H. 3—3.5; Sp.gr. 5.7

In nature arsenic most often occurs as a compound with sulphur and other metals, and forms several minerals. Native arsenic is found only exceptionally, usually in hydrothermal veins, in association with silver, cobalt, and nickel ores. The most important deposits are in Freiberg, Schneeberg, and Marienberg, GDR; Andreasberg and Wittichen, FRG; Jáchymov and Příbram, Czechoslovakia; Schladming, Austria; Baia Sprie, Romania; Kongsberg,Norway; Borgofranco, Italy; Huelva, Spain; Copiapoó, Chile; Hidalgo, Mexico; and in the USA and Australia.

Arsenic (from the Greek *arsenikón*, bold) was isolated in the 4th century, but its compounds were known much earlier, especially its white, deadly-poisonous oxide used by many poisoners, especially at the court of the Byzantine rulers and in Renaissance Venice. At present arsenic is used in metallurgy, in lead hardening (shot), in pharmacy, in dentistry, and in the production of dyes and pigments; its poisonous compounds are used in insecticides.

Allemontite (stibarsen) Sb As
Trigonal; H. 3—4; Sp.gr. 6.2

Allemontite is a natural alloy of arsenic and antimony, macroscopically almost indistinguishable from metallic arsenic. It was identified for the first time in 1822 in the Alpine deposit Allemont, France; and later it was found in ore veins in Andreasberg, FRG; Příbram, Czechoslovakia, and elsewhere.

Crystallized arsenic of cubic, columnar, or platy appearance (3) is comparatively rare; most often it forms fine-grained reniform aggregates with a typical dish- to bulb-shaped separation (1), sometimes fine lath-like grains. In the Japanese deposit at Atakani, Echizen Province, balls of about 1 cm with a prickly surface have been found. On a fresh fracture arsenic is tin-white and of metallic lustre;

when exposed to the air it quickly turns black and loses its lustre. When struck with a hammer, it emits sparks and releases a characteristic garlic odour. The orthorhombic modification of arsenic is the rare mineral called **arsenolamprite** (H. 2; Sp.gr. 5.3—5.5). Allemontite (2) is very similar to arsenic, but is lighter in colour.

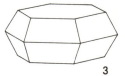

3

2

Bismuth Bi

Trigonal, H.2—2.5; Sp.gr. 9.7—9.8; Melt.point about 280 °C

Bismuth is about the 66th most common element in the Earth's crust, and occurs predominantly in compounds with sulphur. As the native metal it is separated out in mineralization processes accompanying magmatic activity. It is found in pegmatites and, together with bismuthinite and copper minerals, occurs especially in high-temperature quartz veins associated with tin, wolfram, and molybdenum minerals. Deposits of this type are found in Altenberg, GDR; Krupka and Cínovec, Czechoslovakia; Botallack, Cornwall, Great Britain; and Glenn Innes, Chillagoe, and Biggenden, Australia. Large accumulations of bismuth formed in hydrothermal deposits of cobalt, nickel, and silver ores, such as in Schneeberg, Annaberg, and Hasserode, GDR; Wittichen and Schmalkaden, FRG; Jáchymov, Czechoslovakia; Huel Sparnon, Great Britain; Loos, Sweden; St Jean, France; and Temiskaming, Canada. Native bismuth is found in deposits in South America in tin- and silver-bearing ore veins, such as those at Tazna, Illampa, Chorolque, Oruro and Cerro de Pasco.

Bismuth was used in the past in medicine, especially in the manufacture of skin medicaments. Recently it has found an extensive application as a substantial part of alloys with a low melting point which are used in the production of electric fuses, fire detectors, and mould blades (the so-called Wood's metal melts at 60 °C). Bismuth is also used as a semiconductor (its electric resistance changes with pressure), and as a fuel solvent in nuclear reactors.

Perfect crystals of bismuth are rare in nature. Most often it is found in the form of granular aggregates (2). Characteristic skeletal growth gives rise to dendritic, feather-shaped to arborescent forms, flakes and lamellar aggregates (1). It is of pinkish silver-white colour, sometimes with a brass tint, and has a metallic lustre and a lead-grey streak. It is highly diamagnetic, perfectly cleavable, of twinning or longitudinal striation, brittle, and sectile.

1

2

Graphite C
Trigonal; H. 1; Sp.gr. 2.1—2.3; Melt.point above 3,000 °C

Graphite has been used by man since ancient times, and very often was mistaken for the similar molybdenite. Its name is derived from the Greek *graphein,* to write, and was used for the first time by the German geologist, A. G. Werner, in 1789. In 1829 the Swedish scientist Selfström discovered that graphite was a metallic modification of pure carbon, i.e. a mineral of the same chemical composition as diamond.

Graphite is a very common mineral. It originates by the decomposition of compounds of carbon in bituminous rocks and caustobioliths (these are inflammable mineral raw materials of organic origin such as coal and crude oil). From the total amount of 29,000 billion tonnes of carbon found in the Earth's crust at least one-third occurs in the above-mentioned substances, and in the biosphere. In nature most graphite originates by the metamorphism of carbonaceous material of sedimentary origin. It occurs in metamorphosed rocks in all continents.

It is found less frequently in magmatic rocks, e.g. in pegmatites in Ceylon, where it forms a 20-cm thick monomineral filling of numerous veins. It is also found dispersed in volcanic rocks (e.g. in Cornwall, Great Britain, and Harzburg, GDR). Exceptionally it is also found in iron meteorites, e.g. in Canon Diablo, Arizona, USA. In meteorites it also occurs very rarely in the form of cliftonite (a paramorph of graphite after diamond), in deposits in Magura, Czechoslovakia, Toluca, Mexico; and Youndegin, Australia.

Perfect hexagonal tabular crystals with a triangular striation are very rare. Graphite most often forms dark grey to steel-grey aggregates of metallic lustre (1), flakes, radiate leaves, coatings, layers, granular nodules, or masses. It has a black glossy streak, perfect cleavage and flexibility, and is sectile, fire-resistant, and a good electrical and heat conductor.

1

Graphite has been used in the manufacture of pencils since 1550, when the first factory in the world producing pencils was founded in Borrowdale, Great Britain. At present graphite is largely applied in metallurgy, galvanoplastics, and the manufacture of chemical vessels, melting crucibles, lubricants, etc. The latest use of pure graphite is as a moderator to slow down neutrons in atomic reactors.

Diamond C
Cubic; H. 10; Sp.gr. 3.52

Diamond has always been the most valued precious stone. Its name is derived from the Greek *adamas,* invincible. The ancient belief, supported by Pliny the Elder (A.D. 23—79), that diamond was a sort of rock crystal, survived until 1694 when, in a well-known experiment, a diamond specimen was burnt down in Florence in the presence of members of the Italian Academy of Sciences. It was the English chemist H. Davy who proved, in 1814, that chemically diamond is pure carbon.

In nature diamond originates under extreme pressures accompanying, for instance, the volcanic production of kimberlite (a rock composed of peridote, pyroxenes, and pyrope.) The kimberlite fills vertical channels ('pipes') of a diameter of tens or hundreds of metres, which, for instance in South Africa, penetrated rock complexes of a thickness of several kilometres. By the decomposition of kimberlite loose crystals of diamond get into alluvial or placer deposits.

The largest primary deposits of diamonds, discovered in 1871, are situated near Kimberley, South Africa, continuing in a belt to the northeast. The world's largest diamond, which weighed 621.2 g and was called Cullinan, was also found there on 26 January 1905 in the Premier Mine. For several decades these South African diamond deposits produced almost all the world's diamonds. Extensive deposits of a similar type were found in 1961 in Mirny, Yakutia, USSR. Secondary deposits of diamonds occur in Guinea, Liberia, Angola, and elsewhere in Africa, and in Minas Gerais and Bahia, Brazil.

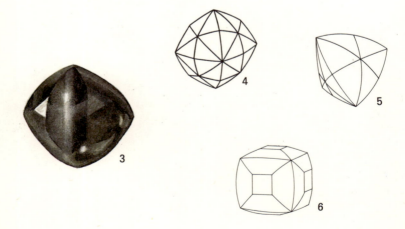

3 4 5 6

Diamond is separated out in the form of isolated crystals (4, 5, 6, 7); it never occurs in granular aggregates. It is the hardest of all minerals but is brittle and fissile; it has adamantine lustre, and is a semiconductor. Usually it is colourless (1), but may also be yellowish, pinkish, brownish (2), greenish, or blue. In ultraviolet light it is luminescent. Its black variety is called bort (3). Because of its physical properties, especially a splendent play of fire in the brilliant cut, diamond is rightly considered the most valuable precious stone as well as a strategic raw material. On account of its hardness bort is used as an abrasive, in rock boring, etc.

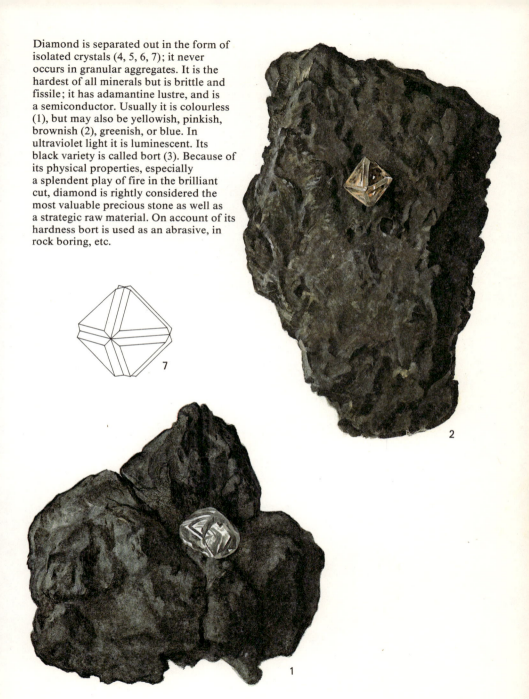

7

2

1

Sulphur S

Orthorhombic or monoclinic; H. 1—2; Sp.gr. 2.0—2.1;
Melt.point about 108 °C

Sulphur was one of the first elements recognized by ancient man from the very beginning of his existence. It is yellow in colour, inflammable, and emits a stinging odour when burned. It was used in Egypt as much as 4,000 years ago for bleaching textiles and as a disinfectant. Since Roman times it has been used in the preparation of medicines. In the Middle Ages supernatural powers were ascribed to sulphur by alchemists. After the 14th century it became an important raw material for the manufacture of gunpowder. As a native element, however, sulphur was identified as late as 1809 by the French chemists Gay-Lussac and L. J. Thénard.

In the list of elements contained in the Earth's crust, sulphur takes up 13th place. A considerable part of it is bound in sulphides or sulphates. Primary native sulphur originates by sublimation as part of volcanic activity, e.g. on Vesuvius, Italy; large deposits are also found near the volcano Popocatépetl, in Mexico, and near Putama, Chile. It also results from the action of hot springs. Much sulphur, however, is of secondary origin. It is the result of weathering of sulphides or of sublimation in fires of coal heaps. The largest accumulations, however, are formed by the reduction of sulphates (gypsum, anhydrite) by special bacteria in the presence of hydrocarbons. Large deposits of this type are in Sicily (Girgenti and Caltanisetta Provinces); other important deposits are in Louisiana and Texas, USA, and at Machów near Tarnobrzeg, Poland. Sulphur occurs there in association with calcite, aragonite, celestite, and barytes.

2

Sulphur forms very brittle, many-faced, perfect or imperfect crystals (4, 5, 6, 7), of yellow to honey-brown colour and adamantine lustre (1). Finely granular and compact, conchoidal, greasy aggregates (2) form large masses, layers, reniform crusts, stalactites, earthy nodules, and granular and powdery efflorescences (3).

Sulphur has become indispensable to modern technology. The largest proportion of it is used in the manufacture of sulphuric acid as a basic material for many industries. Sulphuric acid is also used in the manufacture of cellulose, rubber, dyes and pigments, medicines, disinfectants, and pesticides.

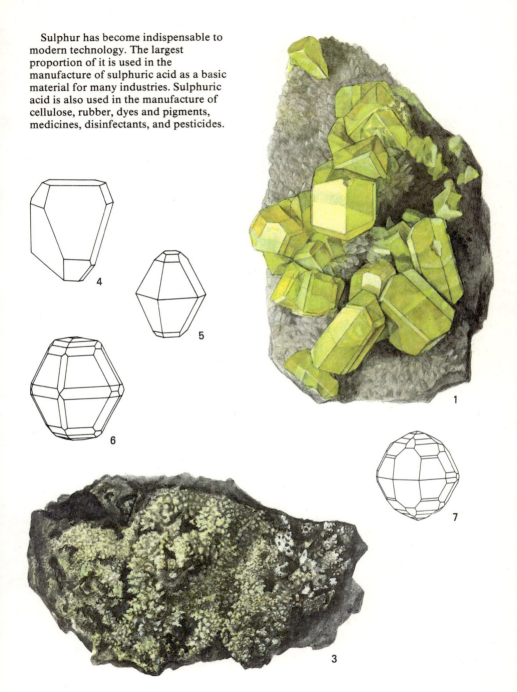

Chalcosine Cu_2S

Monoclinic or orthorhombic (stable below 103 °C);
Hexagonal (stable above 103 °C); H. 2.5—3; Sp.gr. 5.7—5.8

Chalcosine has been known since the 16th century as an important copper ore; (it contains up to 79.8 per cent copper). Usually it is separated out in mineralization processes taking place in deep-seated magmatic zones, but it also occurs in the cementation zone of weathered ore bodies. Chalcosine impregnations and nodules in sediments are of a similar origin. It is found in pegmatites (in California, USA, and Namibia), and in some volcanic rocks where it forms part of large deposits of the 'porphyry copper ores' (in the western part of the USA). In association with chalcopyrite, bornite, tetrahedrite, and enargite it usually occurs in hydrothermal ore veins. The most important deposits are Redruth, Great Britain; Freiberg, GDR (up to 11 per cent silver); Siegerland, FRG; Telemarken, Norway; Vrančice and Bukov, Czechoslovakia; Linares, Spain; Toscana, Italy; the Urals and Altai, USSR; and Butte, Montana and Bisbee, Arizona, USA. Rich deposits are also found in Chile, Zaire (Copperbelt Zone), Transvaal, and near Tsumeb, Namibia. Prospective ores may include mineralized sandstones in the Red Beds, Colorado, USA, and dark impregnated schists near Mansfeld and Sangerhausen, GDR, where chalcosine occasionally also petrifies fossils. Important deposits are found north-west of Wroclaw in Lubin, Glogów, and Legnica, Poland.

Bornite Cu_5FeS_4

Cubic; H. 3; Sp.gr. 4.9—5.3

Bornite occurs associated with chalcosine, and is an important copper ore containing up to 55—69 per cent copper. Important deposits are found in Tarmaya and Los Safros, Chile; in Peru; and Bolivia.

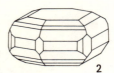

2

1

Crystallized chalcosine in the form of plates (1, 2) or pseudohexagonal twins is rare. It is found mostly at St Just, Great Britain, and Bristol, USA. Compact aggregates are of metal-grey to bluish colour and have a dark blue-grey streak. They are bright, darkening on exposure to the air, and are sometimes coated with a brownish powder.

Bornite aggregates are compact (3) and of conchoidal fracture; on a freshly broken surface they are of copper-red colour (tombac brown), they have a black-grey streak, they tarnish to an iridescent mixture of colours in humid air. Crystals are rare, and are found only in Tincroft, Great Britain; Bristol, USA; and Copiapoó, Chile. Similar minerals are niccolite and pyrrhotine.

3

Argentite Ag$_2$S
Cubic (stable above 179 °C); H. 2; Sp.gr. 7.3; up to 87 per cent Ag

Acanthite
Orthorhombic modification of argentite (stable below 179 °C)

Argentite-acanthite is one of the most important silver minerals and, together with the noble silver ores such as proustite, pyrargyrite, etc., it has contributed to the riches and fame of many world deposits. It occurs in magmatic ore veins of Palaeozoic and Tertiary origin. The primary argentite was separated out from hydrothermal solutions. At lower temperatures it becomes unstable and changes into acanthite. Large amounts of it originate secondarily as a result of the decomposition and replacement of older silver-bearing minerals in the final weathering stage of mineralization. Pseudomorphs after stephanite, proustite, pyrargyrite, native silver, galena, and other minerals are quite common. Some cementation zones, enriched secondarily below the outcrops of ore veins, contain massive argentite (up to 4 kg).

Deposits of argentite-acanthite can be found all over the world. In Europe it is found in Freiberg, Schneeberg, Annaberg, and Johanngeorgenstadt, GDR; Jáchymov, Příbram, Banská Štiavnica, and Měděnec, Czechoslovakia; and Kongsberg, Norway. Comstock Lode, Nevada, USA, is the richest silver-bearing ore vein and is more than 6 km long. It contains compact and earthy argentite and stephanite. Mexico has become the country of silver with 5,000 rich silver veins associated with argentite and other noble silver-bearing minerals. The most important deposits in Mexico are Guanajuato and Zacatecas. In South America argentite-acanthite has been found in deposits in Tres Puntas and Chañarcillo, Chile, and Cerro de Pasco, Peru.

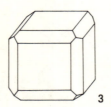

3

Argentite crystals and twins (1, 2), especially octahedral, cubic, rhombododecahedral, and their combinations (3), are usually elongated with uneven faces and rounded edges, and form parallel twins (4) and skeletal bunches. Typical are wires, laminae, plates, compact grains and lumps of different sizes, and black powdery coatings. The original lead-grey colour of argentite turns black in the air and loses its metallic lustre. It has a grey, glossy

4

streak, hackly fracture to indistinct cleavage, is soft, and is perfectly malleable and sectile.

In the past argentite-acanthite was not only important as a source of pure silver. Its perfect malleability also made it especially suitable for coins and it is known to have been used in Freiberg and Jáchymov. Coins minted in Jáchymov were originally called 'Joachimsthaler Gulden' or 'Thaler', from which the name 'dollar' was derived.

1

Sphalerite ZnS
Cubic; H. 3.5—4; Sp.gr. 3.9—4.2; up to 67 per cent Zn

The name of sphalerite was derived from the Greek *sfalerós,* meaning misleading. It is so-called because its colour and unmetallic lustre differ greatly from typical ore minerals. It originates in magmatic, hydrothermal, metamorphic, and sedimentary processes. In association with chalcopyrite, pyrrhotine, magnetite, etc., it forms dark layers in crystalline schists, such as in deposits in Bodenmais, FRG; in the Bohemian and the Saxon Ore Mountains; and in Ammeberg, Scandinavia; Banát, Romania; and Broken Hill, Australia. Large deposits of sphalerite associated with galena originated in the mineralization of hydrothermal veins by metasomatic replacement of limestones and dolomites; examples include Bytom, Poland; Bleiberg, Austria; Westphalia, FRG; Santander, Spain; Belgium; Cumberland and elsewhere in Great Britain; Tri States District, Missouri, Iowa, and Illinois, USA. The deposit in Rammelsberg and the occurrences of sphalerite associated with pyrite and barytes in Meggen, FRG, are probably of sedimentary origin. Usually, however, sphalerite occurs in polymetallic veins of lead and silver ores which are abundant in almost all countries of the world.

Sphalerite had no practical applications in the past, and only became an important raw material for metallurgy in the production of zinc as late as 1860. It contains trace amounts of cadmium, germanium, gallium, and indium, and is an important source of these elements, especially cadmium, for the nuclear industry and for the production of semiconductors.

1

Variously striated, many-faced crystals and twins of sphalerite (1, 4, 5) are usually opaque, translucent or transparent. They have adamantine lustre and are coloured black, brown (3), red, yellow (2), green, or occasionally grey-white or colourless. Sphalerite from Santander has been used in jewellery. Granular aggregates are perfectly fissile; compact and fibrous masses have reniform surfaces, concentrically shelled, of conchoidal fracture and dish-shaped separation. Sphalerite is brittle, is a good electrical insulator, and when rubbed it sometimes phosphoresces.

2

3

4

5

6

7

Wurtzite is a hexagonal modification of sphalerite. It forms dark brown, fan-like, fibrous, resinous, glossy aggregates (6). Tabular and pyramidal crystals (7) are found in Oruro, Bolivia. Small amounts of wurtzite occur in some deposits of sphalerite (e.g. in Příbram, Czechoslovakia).

51

Chalcopyrite CuFeS$_2$
Orthorhombic; H. 3.5—4; Sp.gr. 4.1—4.3; about 35 per cent Cu

Chalcopyrite has been known to miners since ancient times. It is the most common primary copper mineral, found in almost all ore deposits of magmatic or hydrothermal origin. It also occurs associated with sediments deposited in the oxygen-free reduction medium of sedimentary basins. Circulating solutions often change it into malachite, azurite, and some other minerals, and in special conditions also into chalcosite, bornite, and native copper (in the enriched cementation zone). Gold and silver occur as admixtures.

From a great number of deposits only a selection typical of the various kinds of origin will be mentioned. On the large deposit in Sudbury, Canada, chalcopyrite associated with pyrrhotite and pentlandite forms large accumulations in the gabbro and diorite rocks. It occurs dispersed in igneous dykes and forms large deposits of the 'porphyry copper ores' type, mined especially in the USA and South America. Large masses of chalcopyrite are found associated with pyrite in Río Tinto, Spain. Sedimentary chalcopyrite occurs in copper-bearing shales in Mansfeld, GDR; Legnica, Poland; and Rammelsberg, FRG. Usually, however, it occurs, also crystallized, in veins containing ores of non-ferrous metals which are common all over the world.

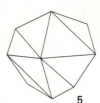

5

2

52

1

Typical chalcopyrite crystals occur in the form of sphenoids (1, 3, 4), many-faced crystals (5), twins, or even individual crystals of distorted shapes. The crystal faces are often striated, of parquet-like pattern, or rough. Aggregates are always compact (2) and of conchoidal fracture. The mineral is typically a characteristic gold-yellow colour tarnishing to green and violet and losing its metallic lustre when exposed to the air. Sometimes a thin black film forms on its surface. It has a greenish grey-black streak, is soft and brittle, and, in contrast to the fragile pyrite, can be marked with a needle, leaving a smooth metallic scratch. Its electrical conductivity increases with rising temperature.

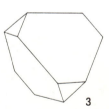

3

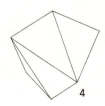

4

Tetrahedrite Cu_3SbS_3

Cubic; H. 3—4; Sp.gr. 4.4—5.4; 30—45 per cent Cu

Tetrahedrite is the name for a whole group of minerals of varying composition, having only an approximate chemical formula. It has been distinguished from other similar minerals since the 16th century and is an important copper ore. It is the product of hydrothermal mineralization and is found in many deposits in association with pyrite, chalcopyrite, bournonite, arsenopyrite, galena, sphalerite, and some silver ores. Its space lattice enables the three main elements to be replaced by others, and therefore it has become the carrier of admixtures. The following varieties of tetrahedrite are distinguished: schwazite, containing up to 17 per cent mercury; freibergite, containing up to 28—36 per cent silver; and annivite, which contains bismuth and usually also iron (2—5 per cent) and zinc (3—6 per cent).

Tetrahedrite is quite a common mineral. The most important deposits are in Freiberg, GDR; Siegerland, FRG; Schwaz, Tyrol, Austria; Pontgibaud, France; Almaden, Spain; St Just, Great Britain; Příbram and Rudňany, Czechoslovakia; Cavnic and Boteş, Romania; Butte, Montana, USA; Huallanca, Peru; Potosí, Bolivia; Copiapoó Chile; and parts of the African 'Copper Belt'. From some of the above-mentioned deposits come unique druses of crystallized tetrahedrite.

2

1

3

Tetrahedrite was named in 1845 after its predominating crystal form — the tetrahedron. Crystals, twins, and penetration twins of different sizes (1) are usually perfectly developed, many-faced (3), and striated; their dull faces often have growths of chalcopyrite and sphalerite crystals. Granular and compact aggregates with uneven to conchoidal fracture (2) are brittle, of metallic lustre, and have a grey- to brown-black streak. Tetrahedrite is of steel-grey or sometimes grey-black colour, with iridescent tarnish.

With an increasing content of arsenic, tetrahedrite changes into **tennantite** (Cu_3AsS_3). Similar minerals are bournonite, stannine, and arsenopyrite. Macroscopic distinction of their aggregates is usually very difficult.

Pyrrhotite FeS
Hexagonal; H. 4; Sp.gr. 4.6

The name of pyrrhotite (or magnetopyrite) was derived from the Greek word *pyrrós*, fire. It is very abundant and is found with many other minerals. Together with pentlandite and chalcopyrite, it was often separated out during the solidification of gabbro-magma; examples are Sudbury, Canada; Bushveld, South Africa; and many deposits in Europe. At high temperatures it originates in ore veins and forms crystals. Deposits of this type are in Freiberg, GDR; Herja, Romania; Leoben, Styria, Austria; and Bottino, Italy. Exceptionally it is transformed from pyrite at extremely high temperatures. As a high-temperature sulphide, pyrrhotite is unstable and often is replaced by pyrite, arsenopyrite, or marcasite, as illustrated by their pseudomorphs which look like its perfect crystals.

Niccolite NiAs
Hexagonal; H. 5.5.; Sp.gr. 7.3—7.8; up to 42 per cent Ni

Niccolite (nickelite, copper nickel, arsenical nickel), very similar to copper ores, was for many centuries considered a worthless mineral, since it did not contain any useful metal. As late as 1751 it was discovered, however, that the mineral is in fact a compound of arsenic and a new, at that time unknown, element — named nickel. Niccolite occurs in hydrothermal veins and belongs to the mineral association of the so-called five-element formation, i.e. the cobalt-nickel-bismuth-silver-uranium formation. There are deposits of the mineral in Jáchymov, Czechoslovakia; Schneeberg, GDR; Wittichen, FRG; and Great Bear Lake, Cobalt City, Canada. It is quite common in ore veins near Mansfeld, GDR; was mined in La Rioja Province, Argentina; and is found with chromite in Los Jarales, Spain.

4

Pyrrhotite crystals are relatively rare, occurring most often in the form of hexagonal plates (2), exceptionally as pyramidal columns with uneven faces (1).

Scale-like crystals are sometimes
arranged in rosettes. Granular or
compact aggregates (3) are of conchoidal
fracture and dish-shaped separation.
Pyrrhotite has a brownish-bronze
(tombac) colour, a metallic lustre, and
a grey-black streak, and is brittle,
magnetic, and a good electrical
conductor. On a fresh fracture it
resembles pentlandite, niccolite, and is
also similar to bornite.

Niccolite rarely occurs in crystallized
form, being most often found in the form
of granular and compact aggregates (4),
and is sometimes even reniform or
arborescent. The copper-red colour turns
brown to grey when exposed to the air,
and loses its metallic lustre. It has a dark
brown streak, imperfect cleavage, and
conchoidal fracture. It is a perfect
electrical conductor. Its surface is often
coated with the pale green annabergite
(nickel bloom), by which it can be
distinguished from bornite or pyrrhotite.

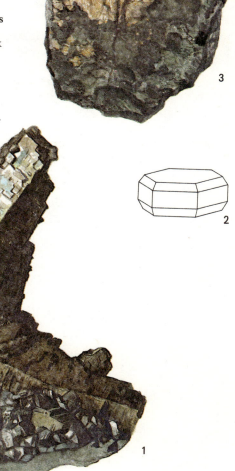

3

2

1

Millerite NiS
Trigonal; H. 3.5; Sp.gr. 5.3; 65 per cent Ni

Small quantities of millerite occur in deposits of different origins. In spite of its high content of nickel it is of no practical application as it is nowhere very abundant. However, it is very beautiful when found in long-fibrous to hair-like bunches filling cavities, and this form is called 'Venus's hairstone'. It was separated out in the cementation zone underlying the primary nickel ores as a product of their decomposition. Its most important deposits are Jáchymov and Příbram, Czechoslovakia; Freiberg and Schneeberg, GDR; Kamsdorf, Riechelsdorf, Bieber and Dillenburg, FRG; Cap Mine, Pennsylvania, USA; and Brompton Lake, Canada. Large accumulations of millerite made up of granular aggregates, crusts, and thick stalactites tens of centimetres long are mined in Agnew, Western Australia. Crystallized millerite often occurs in association with pyrite, chalcopyrite, sphalerite, galena, linnéite, and others, separated out as accessory minerals in the migration of metals in schists and clays, limestone cavities, and in the interiors of pelosiderites in coal deposits. Acicular clusters of particular beauty are found in the coal basin in Kladno, Czechoslovakia; Saarbrücken and Dortmund, FRG; Merthyr Tydfil, Great Britain; and St Louis, Missouri, USA.

Millerite crystals are columnar (2), acicular (up to 8 cm long) and capillary, forming fan-like coatings and spherical structures in cavities (1). It also occurs in irregular and felt-like bunches, finely fibrous, reniform, botryoidal aggregates, or stalactites. It is of metallic to silky lustre, and has a pale brass colour with a greenish tint and a green-black streak. It is brittle, perfectly cleavable and a good electrical conductor. Weathered crystals are often coated with a green carbonate called zaratite.

Galena PbS ✓
Cubic; H. 2.5—3; Sp.gr. 7.2—7.6; up to 86 per cent Pb

Galena has been worked since ancient times, long before it was given its name by the Roman naturalist Pliny the Elder. Like the Babylonians and Egyptians, the Greeks also recovered lead and the associated silver from it (some 20,000 slaves worked in the Lavrion deposit near Athens). The local slags, which were worked later as a valuable raw material by the Romans (and again in the 19th century), contained 6—12 kg of lead and 30 g of silver in 100 kg of old wastes. The Romans also worked deposits in Spain where about 40,000 slaves were employed. In early times consumption of lead was high, since it was used for coins, and in the manufacture of weights, vases, and water piping.

Galena, which is always associated with sphalerite, most often originates from hydrothermal solutions in polymetallic ore veins. Its deposits are found in almost all parts of the world, e.g. in Freiberg and Neudorf, GDR; Stříbro and Banská Štiavnica, Czechoslovakia; Iglesias, Sardinia; Linares, Spain; and in South America and Africa. Galena ore bodies, resulting from the replacement of limestones, are a product of hydrothermal solutions. Deposits of this type are found in Bleiberg, Austria; Bytom, Poland; and Joplin County, Missouri, USA. A deposit of similar origin is found in Broken Hill, New South Wales, Australia. Occasional accumulations of sedimentary origin, metamorphosed at a later date, can be found in Rammelsberg, FRG.

4

3

2

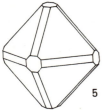

5

1

Galena usually occurs in perfect crystals (1, 2) of varying sizes, predominantly formed by cubes, octahedra, rhombododecahedra, and their combinations (4, 5). It forms granular aggregates, sometimes of linear arrangement, but always perfectly cleavable (3). Compact concentrically banded stalactitic forms up to 10 cm long, with intercalations of sphalerite and marcasite, are called 'tubular ore'. Galena is of lead-grey colour, full or dull metallic lustre, has a grey-black streak, and it is a good semiconductor.

It is the most important lead ore and an important source of silver (up to 1 per cent). In the Middle Ages lead was used in the recovery of silver from other ores, for printing type, and in the manufacture of ammunition, frames for window-panes, and roofing. At present it is mostly used for cable-covers, storage batteries, pigments, and sheets to protect against irradiation.

61

Cinnabar HgS
Trigonal; H. 2—2.5; Sp.gr. 8.1; 86 per cent Hg

Because of its conspicuous red colour, cinnabar attracted man's attention in ancient times, when it was used as a natural dye. Its name, derived from the Old Persian *Kinnavari,* indicates that many ancient localities of cinnabar were found in Asia Minor, e.g. Efésos. The most important deposit, however, was in Sisapon, the present-day Almaden, in Spain. The ore was worked here solely by slaves, who usually died of mercury poisoning after about three years. The Romans were the first to recover native mercury (*argentum vivum*) from cinnabar in clay distillers. Later on, mercury became an important metal in alchemy.

Cinnabar is found in deposits of varied origin. It is the product of mineralization processes taking place at the end of magmatic activity, particularly surface volcanism. It is found at certain thermal hot springs, e.g. Steamboat Springs, Colorado, USA. Cinnabar has been separated out from solutions on ore veins, it has impregnated sedimentary rocks and tuffs, and, having replaced limestones, formed large ore bodies. Associated minerals are pyrite, marcasite, antimonite, and realgar. Powdery crusts originate secondarily in the course of weathering of schwazite (mercuric tetrahedrite).

Cinnabar is quite a common mineral. Important localities are Almaden, Spain (the largest deposit in the world); Idria, Yugoslavia (worked since 1497); Nikitovka, USSR; California, USA; Mexico; and Kvej-Chau, China (where excellent crystals are found).

1

2

Cinnabar occurs relatively rarely in thick-tabular (1, 3) or rhombohedral (4), many-faced, striated, brittle and perfectly cleavable crystals. Granular to compact aggregates (2) are of an uneven to splintery fracture and full submetallic lustre. It also occurs as pulverulent crusts. It has a purple-red or dark brick-red colour, tarnishing to brown or black, and a deep red streak. Cinnabar does not conduct electricity, in contrast to **metacinnabar** (the black cubic modification of HgS, into which it sometimes changes). Similar minerals are realgar, cuprite, proustite, pyrargyrite, some haematites, crocoites, and rutiles.

Cinnabar is the main source of metallic mercury, which was originally used in the recovery of gold and silver from ores, in the manufacture of ammunitions, and in gilding. Mercury is also used in electrochemistry and dentistry. At present its consumption is decreasing.

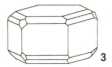

3

4

Antimonite (stibnite) Sb_2S_3
Orthorhombic; H. 2; Sp.gr. 4.6—4.7; 71.4 per cent Sb

Antimonite has been known and used since ancient times; its name has been derived from the Greek *Stíbi*, a silver-grey make-up for darkening the eyelids. The Roman naturalist Pliny described it as a cosmetic preparation much in use in his time. In Japan, its columns of linear crystals up to 1 m long and several centimetres wide, and found in Ichinokawa Mine, Shikoku Island, were used for fencing miniature gardens or to support plants. Large druses of antimonite were used in Japan to decorate homes.

Antimonite occurs in isolated quartz veins and is often gold-bearing. It is also present in hydrothermal ore veins of lead and silver, often in association with cinnabar, orpiment, and realgar. The most important deposits of antimonite, however, originated by the metasomatic replacement of limestones and schists, e.g. in San José del Oro, Mexico; Hamman Meskoutine, Algeria; and Sikvan-Shan, Kiang-Si Province, China. Usually it is found in ore veins, such as in the Auvergne and Vendée, France; Cap Corse, Corsica; Wolfsberg, Goldkronach, GDR; Tuscany; Bohutín, Magurka and Kremnica, Czechoslovakia; Cavnic and Săcărâmb, Romania; and Nikitovka, USSR. Large and brilliant crystals have been found in Kostajnik, Yugoslavia; Pereta and Cetina, Italy; and Lubilhac, France.

4

3

Crystals of antimonite are always unidirectionally developed, narrowly columnar (1, 2, 4), needle-shaped to capillary, many-faced (5), pointed-pyramidal (6), and with longitudinal striation. Aggregates are usually rod-like (3), fibrous, or felt-like, sometimes radiate in bunches, and exceptionally are compact with conchoidal fracture. Antimonite has a grey streak, is of lead- to steel-grey

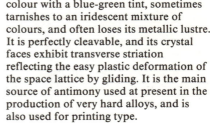

colour with a blue-green tint, sometimes tarnishes to an iridescent mixture of colours, and often loses its metallic lustre. It is perfectly cleavable, and its crystal faces exhibit transverse striation reflecting the easy plastic deformation of the space lattice by gliding. It is the main source of antimony used at present in the production of very hard alloys, and is also used for printing type.

65

Tellurides of gold and silver

Tellurium is one of the few elements combined in nature with the comparatively inert gold, and forms commercial ore bodies of tellurides. The telluride group includes several minerals, the most important being:

Calaverite $AuTe_2$. Monoclinic; H. 2.5; Sp.gr. 9; up to 44 per cent Au.
Krennerite $(Au,Ag)Te_2$. Orthorhombic; H. 2—3; Sp.gr. 8.5;
   30—40 per cent Au.
Sylvanite $AuAgTe_4$. Monoclinic; H. 1.5—2; Sp.gr. 8.0—8.3;
   25—27 per cent Au.
Petzite $(Ag,Au)_2Te$. Cubic; H. 2.5; Sp.gr. 8.2—9.4;
   up to 25 per cent Au.
Nagyagite $AuPb_5,(Sb,Te)_4S_6$. Orthorhombic; H. 1—1.5;
   Sp.gr. 6.8—7.5; 6—13 per cent Au.
Hessite Ag_2Te. Monoclinic; H. 2.5; Sp.gr. 8.3—9.0;
   up to 63 per cent Ag.

All of these minerals are separated exclusively from hydrothermal solutions either on ore veins or in propylites produced by the Tertiary volcanism, or in quartz veins formed by earlier magmatic intrusions. They form aggregates and druses composed of tabular or columnar crystals. The tellurides sometimes become coated with dark brown powdery to spongy encrustations containing native gold.

The most important localities of these minerals are in Săcărâmb, Baia de Arieş, Boteş, and Zalatna, Romania; Calaveras County, California, Cripple Creek District and Boulder County, Colorado, and Huronian Mine, Ontario, USA. In Western Australia the most important deposit is in Kalgoorlie.

2

5

1

Sylvanite is called 'letter ore' after the typical groups of flattened linear skeletic crystals (1,4). It is of silver-white colour with a yellowish tint, of metallic lustre, and is perfectly cleavable.

Nagyagite only exceptionally crystallizes in the form of flexible plates (5); most often it occurs in scaly, lead-grey aggregates of metallic lustre and perfect cleavage. It is called 'lamellar ore' (2).

Hessite is also a metallic grey colour with a bluish tint, is non-cleavable, and is partly sectile. Cube-shaped crystals often form clusters (3).

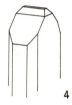

4

3

Pyrite FeS$_2$ ✓
Cubic; H. 6—6.5; Sp.gr. 5.0—5.2; up to 53 per cent S

Apart from quartz and limestone, pyrite is one of the most common minerals and the most abundant ore-bearing mineral. It has been known since ancient times, and its name is derived from the Greek *pyrites*, meaning sparkling (sparks fly off when it is struck). It was believed to possess healing properties, and the Incas polished it to mirrors, but its real importance was discovered as late as the Middle Ages by alchemists. Pyrite originates during mineral-forming processes of almost all types. It gets separated from the solidifying magma and often forms inclusions in many igneous rocks. It is quite common in hydrothermal ore veins and genetically similar metasomatic accumulations. It contains trace amounts of gold and copper. It is also found as a petrifaction medium in sedimentary rocks, e.g. in bituminous limestones, argillaceous slates, schistose clays, and coal seams. Stratiform ore deposits in crystalline schists originate through metamorphosis. Sedimentary pyrite weathers into what is called vitriol (alum) shales, which have been used in Bohemia since the 16th century in the production of the so-called fuming sulphuric acid (*olea vitrioli*).

The chief producers of pyrite in Europe are Río Tinto, Spain; Meggen, FRG; Smolník, Czechoslovakia; and Falun, Svealand, Sweden. The best known crystals come from Elba, Traversella, and Brosso, Italy; Gellivare, Sweden; Central City, Colorado, USA; and Agua Blanca, Mexico.

The most common forms of pyrite crystals are striated cubes (2), pentagondodecahedra (1,4,5), octahedra, and their combinations. Granular aggregates, radiate-fibrous crusts, and loose nodules are often reniform. The mineral is brass-yellow and sometimes displays an iridescent tarnish (3). It has a brown-black streak, uneven fracture, metallic lustre, and compact aggregates are often dull. Pyrite aggregates are similar to those of marcasite; acicular crystals resemble millerite.

Pyrite is used in the manufacture of paints, varnishes, blue and green vitriol (copperas), and fine polishes. In the past it was also used for making jewels. The largest quantities nowadays are used in the manufacture of sulphuric acid, which

1

is an important agent for the production of fertilizers, plastics, medicines, and explosives, and is also used in working ores and crude oil, and in electrochemistry.

2

5

4

3

Marcasite FeS$_2$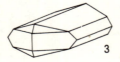

Orthorhombic; H. 5.5—6; Sp.gr. 5.9—6.2

Marcasite has been known since ancient times and was used in the same way as the similar pyrite. However, it was only determined as a mineral as late as the 19th century. Marcasite is not as abundant as pyrite. It is separated out from low-temperature acid magma solutions close to the surface of the Earth and only part of it is found in deep-seated magmatic rocks. In hydrothermal veins it is usually the youngest mineral. It also originates from the decomposition of pyrrhotite and forms pseudomorphs after pyrrhotite crystals. Large masses of marcasite are found in some coal seams and other sedimentary rocks, as the remains of vegetable and animal bodies contained in them supplied sufficient amounts of sulphur to form marcasite. Workable deposits were formed by the metasomatic replacement of limestones and dolomites.

Crystal druses of marcasite are found in ore veins, e.g. in Freiberg, GDR; Clausthal, FRG; Pontpéan, France; and in Romania and Great Britain. It is also present in polymetallic deposits near Aachen, FRG; Bleiberg, Austria; Bytom , Poland; and Wisconsin, USA. Perfect crystals, twins, concretions, and globular nodules occur in limestones and marls of the Cretaceous Period on the shores of the Channel in Folkestone and Dover, England, and Boulonnais, Calais, and in many basins in France. Isolated crystals of large size and an excellent quality are found in clays in the brown-coal basin in Most and Duchcov, and in the Vintířov Mine near Sokolov, Czechoslovakia.

Columnar and tabular (3), embedded as well as attached, striated crystals of marcasite usually form twins or hypoparallel jagged aggregates called 'pectiniform pyrite' and 'spear pyrite' (1,4). Radiate, fibrous, and compact aggregates separated out in the form of granules, reniform coatings and crusts, stalactites or irregular nodules or radiate-fibrous isolated globules (2).

Marcasite is of bronze-yellow colour with green or iridescent tarnish, and has a dark grey-green streak, uneven fracture, metallic lustre, and is brittle. In damp air it weathers more easily than pyrite, producing sulphuric acid which often causes chemical decomposition of other minerals. Both pyrite and marcasite are mainly used in the chemical industry.

4

1

2

Arsenopyrite FeAsS
Monoclinic (pseudorhombic); H. 5.5—6; Sp.gr. 5.9—6.2

Arsenopyrite is a typical lode-ore mineral separated out in magmatic mineralization processes. It is quite common in veins of gold-bearing quartz, belongs to the family of tin, wolfram, and molybdenum ores, and is present in accumulations of silver, nickel, and cobalt ores, as well as in deposits of siderite and chalcopyrite. It was also separated out in ore bodies of metasomatic origin and occurs disseminated, or forms intercalations and large encrustations, in crystalline schists.

Arsenopyrite druses are found at St Just, Great Britain; Panasqueira, Portugal; and in the Saxon and Bohemian Ore Mountains. Perfect crystals are found in hydrothermal veins in Freiberg, GDR; Andreasberg, FRG; Příbram, Czechoslovakia; Mitterberg, Austria; Dépt. Haute-Vienne and Ariège, France; Tanza, Bolivia; Coquimbo-Buitre, Copiapoó, Chile; and Goulbury, Australia. Massive aggregates are found in Złoty Stok, Poland. It is an important mineral of quartz veins in many countries because of its content of disseminated gold. It is also mined for its cobalt content in the large deposits in Skutterud, Sala, and Boliden, Scandinavia. Arsenopyrite devoid of its valuable admixtures is extensively used in the production of insecticides, and as a source of arsenic which is used as alloy addition in metallurgy.

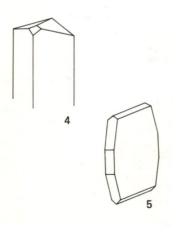

4

5

Arsenopyrite forms short- (2, 3) or long-columnar (4), tabular (5), striated crystals of orthorhombic profile, embedded, isolated, or attached on the walls of cavities (1). The crystals are often intergrown. Aggregates are usually massive, granular, rod-like to fibrous, of uneven fracture, and form granules, nodules, reniform crusts, and exceptionally also interlacing masses. Arsenopyrite is tin-white to steel-grey in colour with a yellowish tint, is sometimes of iridescent tarnish, and has a metallic lustre. When struck it emits sparks and a characteristic garlic odour. In some deposits it weathers easily and becomes covered with greenish decomposition products (e.g. scorodite and pitticite). Similar minerals are löllingite, skutterudite, and the softer gudmundite.

1

3

2

Molybdenite MoS_2

Hexagonal; H. 1—1.5; Sp.gr. 4.7—4.8; 59.9 per cent Mo

Molybdenite was often in the past mistaken for graphite as the two minerals are similar in appearance as well as properties. Molybdenite was identified as a separate mineral as late as the 18th century. It is a typical product of mineralization processes combined with intrusions of very acid igneous rocks. It is quite common but mostly occurs in small amounts, e.g. disseminated in granite, filling its cracks, or in pegmatite veins. Workable deposits of molybdenite originate from high-temperature solutions, together with cassiterite and wolframite, in greisens (metamorphosed granites) and in quartz veins of the characteristic mineral family of tin, wolfram, and molybdenum ores. Molybdenite often occurs as a component of ore deposits at the contact with granite bodies. Disseminated molybdenite can also be found in dark copper-bearing shales near Mansfeld, GDR, and in south-west Poland. Rich deposits are found in Climax, Colorado, USA; Chillagoe and Herberton Districts, Queensland, Australia; and Stavanger and Flekkefjord, south Norway. These three countries are responsible for the major part of the world production of molybdenite. Smaller deposits are found in the Saxon and Bohemian Ore Mountains; Cornwall, Great Britain; Spain; Canada; Peru; and some other countries.

Molybdenite crystals are very rare, although excellent specimens have been found in Slangsvold and Moos, Scandinavia; Renfrew, Canada (crystals weighing up to 0.5 kg); Edison, New Jersey, USA, and New South Wales, Australia.

1

2

3

Hexagonal crystals in the form of thick plates (1,3) and pyramidal columns are very rare. Molybdenite most often forms scaly aggregates (2), and flexible laminae, sometimes with a diameter of up to 30 cm (a typical deposit is found at Quebec, Canada). Molybdenite is lead-blue in colour with a violet tint, has a dark blue-grey streak (with a greenish tint on porcelain), and a metallic lustre, and is perfectly cleavable, soft, and sectile. Because of its high melting point it is used as a lubricant. Molybdenite resembles lead and, in fact, its name comes from the Greek *molybdos,* meaning lead.

It is the most important molybdenum ore, and is used as an alloy addition in special steels. It contains workable quantities of rhenium which is one of the rare elements.

Skutterudite $CoAs_{2-3}$
Cubic; H. 5—6: Sp.gr. 6.4—6.9; up to 28 per cent Co

Nickelskutterudite $NiAs_{2-3}$
Cubic; H. 5—6; Sp.gr. 6.4—6.9; up to 28 per cent Ni

The group of cobalt and nickel arsenides contains minerals which were originally called smaltite and chloanthite. Both had varying contents and consequently unstable proportions of cobalt, nickel, arsenic, and iron. Later they were called skutterudites after the well-known deposit in Skutterud, Norway.

Both these minerals are found in hydrothermal ore veins of the five-element formation of nickel-cobalt-bismuth-silver-uranium. Associated minerals are niccolite, safflorite, rammelsbergite, and maucherite. This type of deposit is found in Schneeberg and Annaberg, GDR; Andreasberg, Wittichen, and Riechelsdorf, FRG; Jáchymov, Příbram, Horní Slavkov, and Měděnec, Czechoslovakia; Chalanches, France; Loos, Tunaberg, Sweden; Skutterud, Snarum and Modum, Norway; Great Bear Lake, Temiskaming and Cobalt City, Canada; and Australia. Large crystals come from Bou-Azzer, Algeria.

3

1

2

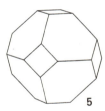

5

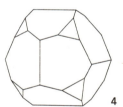

4

Crystals of skutterudite and nickelskutterudite (3, 4, 5) are rare. They are octahedral, pentagondodecahedral (1), and cubic, often of irregular shape, with step-like, slightly convex faces. Aggregates are usually compact (2), granular, or reticulated, sometimes reniform, of uneven fracture, and grey-black streak. Skutterudite is tin-white to steel-grey (1), of metallic lustre, and tarnishes to pink (nickelskutterudite (2) tarnishes to a greenish colour). When struck the mineral emits a garlic odour. Skutterudite is readily affected by weathering and becomes coated with efflorescent crusts of pink erythrite; nickelskutterudite becomes coated with green annabergite (2). Similar minerals are arsenopyrite, löllingite, safflorite and rammelsbergite. Skutterudite and nickelskutterudite are important ores of cobalt and nickel. Both these elements are used in metallurgy, electronics, electrochemistry, glass-making, and elsewhere.

Pyrargyrite Ag$_3$SbS$_3$
Trigonal; H. 2.5—3; Sp.gr. 5.8; 60 per cent Ag

Proustite Ag$_3$AsS$_3$
Trigonal; H. 2.5; Sp.gr. 5.5; 65 per cent Ag

The group known as noble silver ores contains some chemically related minerals, compounds of silver and sulphur with antimony or arsenic. The most common are pyrargyrite and proustite. They only occur together in hydrothermal veins in deposits of polymetallic lead and zinc ores. Proustite prevails in vein deposits of the five-element formation — nickel-cobalt-bismuth-silver-uranium with arsenic. Associated minerals include galena, sphalerite, skutterudite, and some other rare minerals of cobalt and nickel, usually stephanite, polybasite and other rare silver-bearing minerals also occur. Exceptionally, **pyrostilpnite** and **xanthocone**, monoclinic minerals with compositions adequate to pyrargyrite and proustite, are present.

Pyrargyrite and proustite are found in many deposits, e.g. Freiberg and Marienberg, GDR; Wittichen, Andreasberg, and Markirch, FRG; Jáchymov, Příbram, and Banská Štiavnica, Czechoslovakia; Chalanches, France; and Hiendelaencina, Spain. Both these minerals are also important silver ores on the American continent. They are found in Arizona and Colorado, USA; in the area of Guanajuato and Zacatecas, Mexico; in Bolivia; and in Peru. Perfect large crystals come from the well-known deposit in Chañarcillo, Chile.

Pyrargyrite forms many-faced perfect columnar (3), sometimes striated crystals, filling cavities in veins (2). Proustite crystals have almost identical appearance (1); they are, however, not so many-faced. Granular to compact aggregates of splintery fracture form granules, crusts, and dendritic encrustations, and, exceptionally, pseudomorphs after stephanite. The typical original cinnabar-red colour of proustite and the dark red colour of pyrargyrite become dark when exposed to various forms of radiation, and the originally translucent minerals become almost black. Their specimens in collections should therefore be protected from the light. Both minerals have a red streak, which is lighter in proustite — 'light red silver ore' — than in pyrargyrite — 'dark red silver ore'. Aggregates of cinnabar, cuprite, miargyrite and haematite are similar in appearance.

1

2

Boulangerite $Pb_5Sb_4S_{11}$
Monoclinic; H. 2.5; Sp.gr. 5.8—6.2; 55—58 per cent Pb

Boulangerite is named after the French mining engineer C. L. Boulanger. Sometimes it is also called plumosite. It is one of the so-called felt-like ores forming irregularly interlaced fibrous accumulations. This group also includes jamesonite, zinckenite, and meneghinite. Felt-like ores are compounds of lead and antimony with sulphur, and chemically can be distinguished only by the varying proportions of these elements.

Boulangerite is the product of hydrothermal mineralization. A fairly rare mineral, it originates in polymetallic ore veins from solutions which probably have a low concentration of sulphur. Associated minerals are galena, sphalerite, antimonite, and tetrahedrite; the gangue is predominantly composed of quartz and siderite. Boulangerite is of no commercial importance since it usually forms only admixtures in the worked ores. Furthermore, it occurs only in a few deposits, e.g. Wolfsberg, GDR; Oberlahr, Wissen, FRG; Schneeberg, Austria; Trepča, Yugoslavia; Baia Sprie, Romania; Sala, Boliden, Sweden; Nasafjord, Norway; and Nertschinsk, Transbaikal, USSR. However, large accumulations of boulangerite were mined in Molière, France. It is also found disseminated in the grey quartz ('Dürrerz') in Příbram, Czechoslovakia.

3

Boulangerite, as has been said before, is neither common nor commercially important, although some crystallized specimens are fancifully shaped. Bunches of longitudinally striated needles and capillaries (several centimetres long) are found in vein cavities (1); fine filaments are often interlaced in feather-like bunches. More frequent are acicular disseminated crystals and finely fibrous to compact aggregates (2), of dark lead-grey colour and dull, silky lustre.

Fibrous, wedge-shaped crystals are
flexible, unlike the similar crystals of
jamesonite (3 — view of top part), which
are brittle and have a transcrystalline
fracture. A similar mineral is antimonite.

Realgar AsS
Monoclinic; H. 1.5—2; Sp.gr. 3.5—3.6; up to 70 per cent As

Realgar has a conspicuous red colour and because of this it was often mistaken for certain ore minerals by miners in the past. Its chemical composition was determined as late as 1810. It is not very common. It usually originates associated with orpiment from low-temperature solutions resulting from volcanic activity and the related issues of hot thermal springs. It forms secondarily in ore deposits as a product of weathering of arsenic ores, and often is the distillation product in burning coal piles. In the past it was used as a paint; at present it is sometimes utilized as a source of arsenic for industrial purposes.

Important deposits of realgar are found in Cavnic and Săcărâmb, Romania; Alshar, Macedonia; Getchel, Nevada, USA; Persia and China. Attractive crystals are found in Binnental, Wallis, Switzerland, and encrustations form on Vesuvius, Italy, and in Yellowstone, USA.

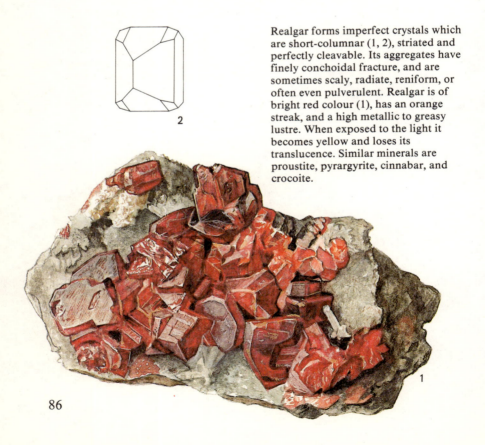

2

Realgar forms imperfect crystals which are short-columnar (1, 2), striated and perfectly cleavable. Its aggregates have finely conchoidal fracture, and are sometimes scaly, radiate, reniform, or often even pulverulent. Realgar is of bright red colour (1), has an orange streak, and a high metallic to greasy lustre. When exposed to the light it becomes yellow and loses its translucence. Similar minerals are proustite, pyrargyrite, cinnabar, and crocoite.

1

Orpiment As$_2$S$_3$

Monoclinic; H. 1.5—2; Sp.gr. 3.4—3.5; up to 61 per cent As

Orpiment's name is due to its gold-yellow colour, and the Ancient Romans believed it to be an auriferous ore. For many centuries it was imported from the East to be used in paints. It originates in the same way as realgar, and often is the product of the former's metamorphosis. Consequently, both minerals occur in association. The most important deposits are Julamark, Kurdistan; and Acobambilla, Peru, where large accumulations of long-bladed, petal-like, tabular aggregates are found. Orpiment is used for the same purposes as realgar.

2

Perfectly crystallized orpiment (2) is rare. Irregular, short-columnar, striated crystals are usually clustered in spherical aggregates. Its scaly, thick tabular (1) and radiate-fibrous aggregates are perfectly cleavable; reniform aggregates are of dish-shaped separation. It is also pulverulent, of yellow colour, has a yellow streak, is of pearly to greasy lustre, and is translucent, sectile and flexible.

1

Halite (common rock salt) NaCl
Cubic; H. 2; Sp.gr. 2.1—2.2; about 39 per cent Na and 60 per cent Cl

Since the early history of man rock salt has played an important part. While other mineral raw materials have been used since early ':mes to increase the technical advancement of human society, rock salt is indispensable for human life itself (the average annual consumption per head is 7 kg). Therefore, it acquired a considerable commercial importance, and was much in demand along the so-called salt paths.

Rock salt is one of the most common minerals in nature. Large amounts of rock salt are present in the waters of oceans (the average salinity of sea water is 3.75 per cent, of which more than three-quarters is NaCl) and salt lakes. Extensive deposits of rock salt were formed by the evaporation of sea water in shallow basins, sometimes associated with gypsum, anhydrite, or potassium salt deposition. It became a rock-forming constituent in the geological structure of vast territories. Salt impregnates some clays and marls, and in desert or steppe areas forms thin encrustations or efflorescences. Halite may also be of volcanic origin (e.g. as sublimate on Vesuvius and some other volcanoes).

Large accumulations are found in Palaeozoic to Tertiary sediments from all over the world. In Europe large deposits are found near Stassfurt, GDR; Würtemberg, FRG; Salzkammergut, Austria; Wieliczka, Poland; and in Romania and Great Britain. Karst phenomena are known from Cardona, near Barcelona, Spain, and large crystals of up to 30 cm come from California, USA.

3

Halite crystallizes in regular or irregular cubes (1). Aggregates are usually compact, granular, fibrous (3), and of conchoidal fracture. Hair-like forms, skeletal efflorescences and stalactites are quite common. Halite is colourless and translucent to transparent. Its bluish to

violet tint (2) is caused by scattered metallic sodium; red salt is coloured by haematite inclusions (3). Clay and the bitumens colour the salt grey-black. Halite has a perfect cleavage, glassy lustre, and is hygroscopic when it contains an admixture of calcium and magnesium.

Halite is an indispensable mineral in the food and preserving industries, and the most important raw material in the chemical industry in the production of sodium, chlorine, and their compounds. Large pure crystals are used in optical apparatus as the mineral transmits infrared rays.

Fluorite CaF_2

Cubic; H. 4; Sp.gr. 3.1—3.2; about 48 per cent F

Fluorite has been a popular mineral since ancient times. In Greece and Ancient Rome it was regarded as a precious stone and was used in the manufacture of various articles of artistic value, e.g. vases (*vasae murrhinae*). It is widely distributed (but seldom abundant), mainly in regions penetrated by granite, syenite, and rhyolite massifs. As part of volatile compounds escaping from the acid magma, it exceptionally originated in igneous rocks and in pegmatites, but predominantly in greisens in association with topaz, tourmaline, apatite, and lithium micas. It also occurs in quartz veins associated with tin, wolfram, and molybdenum ores, as well as in deposits of lead and copper ores. Fluorite associated with barytes also forms low-temperature veins and impregnates sandstones. Fine crystals are found in the druses of quartz, albite, adularia, rutile, and other minerals of the so-called Alpine paragenesis on cracks in crystalline schists, e.g. in Switzerland and Austria. Extensive metasomatic deposits of fluorite originated through the replacement of limestones, e.g. in Kentucky, Illinois, USA, and Las Cuevas, Mexico.

Fluorite is found almost throughout the world; the most important deposits are in Alston, Weardale, Great Britain (deep violet coloured crystals); Kongsberg, Norway, and Transbaikal, USSR (green); Freiberg, Annaberg, GDR (yellow); St Gotthard, Chamonix, the Alps (pink); Strzegom, Poland; and Nocera, near Naples, Italy (colourless). When the violet-black variety of fluorite with bitumen is struck, it emits a bituminous odour. It is found in Moldava, Czechoslovakia.

3

4

90

5

6

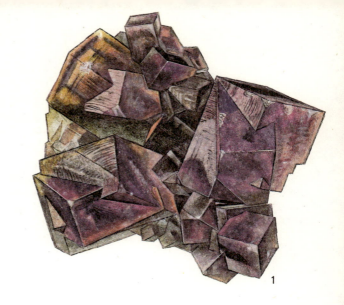

1

Fluorite crystallizes in the form of cubes (2), octahedra, rhombododecahedra, their combinations (4, 5), and twins (1, 6). Aggregates are usually coarsely granular, compact, and sometimes of sprudelstein texture (3). Different colour shades in an individual crystal are partly due to the content of trace elements of rare earths, partly to different defects in the crystal lattice. Fluorite is perfectly cleavable, brittle, translucent to transparent, and of a full to dull lustre. It is the most important raw material in the production of fluorine and its compounds for the chemical, glass and nuclear industries. In the past large amounts of fluorite were used in metallurgy. Limpid crystals transmitting ultraviolet rays are used in optics.

2

91

Cuprite Cu_2O

Cubic; H. 3.5—4; Sp.gr. 5.8—6.2; about 88 per cent Cu

This mineral's name was derived from *cuprum,* meaning copper, in 1845. However, it had been known to miners for many centuries before as a copper-bearing mineral. It originates secondarily by the oxidation of various primary copper ores, especially sulphides. Consequently it forms mainly on exposures in the cementation zone of ore accumulations. Associated minerals usually include native copper, malachite, azurite, and other secondary copper-bearing minerals. It often forms pseudomorphs after crystals and wires of native copper, and, being unstable, it also often changes partly or completely into pseudomorphs of malachite. The earthy up to compact admixture of cuprite and limonite is called 'brick ore' and originates especially by weathering of chalcopyrite. Cuprite of magmatic origin was found in 1835 in volcanic bombs on Vesuvius, Italy.

Cuprite is a very common mineral, with a high content of copper, but only exceptionally forms commercially important deposits. Perfect crystals several centimetres across are found at Liskeard, Cornwall, Great Britain; Chessy, France; Moldova, Romania; and Gumeshevsk and Mednorudjansk Mines and Nizhnii Tagil, Central Urals, USSR. Outside Europe the most important deposits are at Tsumeb, Namibia; Bisbee, Arizona, USA; Chile; Peru; and South Australia.

Cuprite crystals occur in the form of octahedra and rhombododecahedra (2) in druses (1); sometimes they are embedded, isolated, and of perfect growth in all directions. The fine hair-like variety is called **chalcotrichite**. Perfect specimens are found in Rheinbreitbach, FRG.
Cuprite aggregates are granular, compact,

and earthy. Cuprite is dark crimson in colour but tarnishes in the air to golden or grey, has a red to red-brown streak, is translucent and glossy, and its aggregates have a grey, metallic lustre. Transparent crystals are very rare and are classed as precious stones (e.g. from Onganya, Namibia, and from several new locations in Pennsylvania, USA). They are much prized today for their wonderful colour and high adamantine lustre. The low hardness of cuprite, however, is the reason why it cannot be used cut in jewellery. It is a much-valued gem for collections and its price rises tremendously with the size of the crystal. Similar minerals are proustite, pyrargyrite, cinnabar and haematite. Malachite crusts may serve as a distinguishing feature.

1

Magnetite Fe_3O_4
Cubic; H. 5.5; Sp.gr. 5—5.2; Melt.point about 1,500 °C; up to 72 per cent Fe

---

Because of its magnetism, magnetite attracted the attention of many natural scientists in Ancient Greece and Rome. According to preserved records, however, it was already known to the Chinese as early as the 11th century B.C. It is a widely distributed mineral, usually occurring disseminated in volcanic rocks and crystalline schists. During magmatic differentiation it is concentrated in masses of dark as well as light igneous rocks (gabbros and syenites). It also originates by the metamorphosis of haematite and limonite deposits, and accumulates in skarns (metamorphic rocks associated with garnet, hedenbergite, amphibole, and epidote). Magnetite crystals occur in hydrothermal veins and in cracks of crystalline schists in association with minerals of the so-called Alpine paragenesis. Only exceptionally does it originate from the sedimentation of iron ores. It is found in workable quantities in alluvial deposits.

The most important deposits of magnetite are in Scandinavia; the Urals, USSR, the USA, and Canada. In Europe it is mined in Kirunavaara, Gellivare, and Långbån, Sweden; Arendal and Sydvaranger, Norway; and Magnitnaja Gora and Blagodat, Urals, USSR. Perfect crystals are found in Binnental, Switzerland; Moravița, Romania; and Traversella, Italy.

Chromite $FeCr_2O_4$
Cubic; H. 5.5; Sp.gr. 4.5—4.8; 52—58 per cent Cr_2O_3

Chromite is the only important chromium ore. It is the product of magmatic differentiation, and as such is found in ultrabasic rocks (peridotites and dunites) and serpentines formed by their metamorphosis. The most important deposits of chromite are in Transvaal, South Africa; Selukwe, Namibia; Brussa, Turkey; Albania; Norway; and New Caledonia.

3

4

1

Magnetite crystals in the form of striated octahedra (1, 3) are mostly embedded, attached, or finely disseminated, sometimes of irregular growth or in the form of twins (4). Granular as well as compact aggregates have conchoidal fracture. Magnetite is black, opaque, of metallic lustre, strongly magnetic, and has a black streak. It is one of the most valuable ores of iron. Similar minerals are chromite, ilmenite, and hausmannite.

Crystallized chromite is fairly rare. Compact black to brown-black aggregates have a metallic to greasy lustre and a brown streak, and form granular nodules, bunches, and tabular layers (2). Chromite is used in the production of paints, in chemistry, in glass-making, and especially in the production of chromium for special steels (ferrochromium) and as a non-rusting plating of metals. Similar minerals are magnetite and franklinite.

2

95

Valentinite Sb₂O₃

Orthorhombic; H. 2.5; Sp.gr. 5.7; up to 83 per cent Sb

Valentinite is one of the minerals originating by the decomposition of primary antimony ores. It was first described at the close of the 18th century in Allemont, France, and shortly afterwards was also found in the upper parts of polymetallic veins in Příbram, Czechoslovakia, where the brilliant crystals up to 3 cm long are found. In the past valentinite was bought by pure gold in amounts equalling the weight of the crystals; one large crystal is said to have cost one or two golden ducats.

Valentinite is not very common. It occurs only in some antimonite deposits, and is of no practical significance. The most important localities are in Bräunsdorf, FRG; Wolfsberg, GDR; Pezinok-Pernek, Czechoslovakia; and Baia Mare, Romania. Accompanied by **senarmontite** (cubic; Sb₂O₃), it is found in large quantities in Sanza, Algeria, and Tatasi, Bolivia. It was mined in these deposits as an important antimony ore.

Valentinite usually occurs in the form of columns, less often as plates (2), lenticles, or irregular clusters. It forms radiate-fibrous aggregates in fissures (1), and also pseudomorphs, mostly after antimonite. It is colourless, milk-white, or yellowish, translucent, perfectly cleavable, soft, brittle, and easily friable (it breaks easily). It has adamantine, silky, or pearly lustre. Cerussite is a similar mineral.

2

1

Kermesite Sb_2S_2O
Monoclinic; H. 1—1.5; Sp.gr. 4.5

Kermesite, together with powdery antimony ochres, valentinite and senarmontite, results from weathered antimonite. It is not very common, and is of no practical use. Bunches of acicular crystals are very attractive. Perfect crystals are found in Pezinok-Pernek, Czechoslovakia; Pereta, Tuscany, Italy; Bräunsdorf, near Freiburg, FRG; and Djebel Haminat, Algeria.

In the interior of cavities kermesite forms radiate clusters of needle-shaped crystals and bunches thickly set with hairy crystals several centimetres long. In fissures it forms radiate-fibrous, stellate crusts (1). It is cherry-red, with a high or silky lustre, and is translucent, soft, brittle, and perfectly cleavable. Similar minerals are chalcotrichite and xanthocone.

1

97

Corundum Al_2O_3
Trigonal; H. 9; Sp.gr. 3.9—4.1

Corundum is a commonly found mineral originating in the course of various rock-forming processes in the presence of a large amount of aluminium. It therefore mostly occurs in igneous rocks (gabbros and syenite), in granite pegmatites, in crystalline schists (gneisses), and in contact metamorphic limestones and dolomites. From these mother rocks, it has also penetrated alluvial deposits. The fine-grained mixture of corundum, magnetite, haematite, and quartz, is known as emery and is used as an abrasive. It is mined on Naxos Island, Greece; near Smyrna, Turkey; and in Chester, USA.

Corundum is widely distributed but never abundant. In Europe it is found in Routivara, Sweden; Pokojovice, Czechoslovakia; St Gotthard, Switzerland; and Mias, Southern Urals, USSR. Important deposits are in Georgia, USA; Renfrew, Canada; Pietersburg, Transvaal, South Africa; and Chittering, Western Australia.

The two precious varieties of corundum — ruby and sapphire — were known long before common corundum. Both these precious stones have been mined since ancient times from alluvial deposits and crystalline limestones in Ratnapura, Ceylon; India; Madagascar; Mogok, Burma; and Thailand. Ruby and sapphire deposits were recently also discovered in Pailin, Cambodia; Borneo; North Carolina, USA; Africa; and Australia. The best sapphires in Europe come from Jizerská louka, Czechoslovakia; new deposits of rubies were recently found in Prilep, Yugoslavia.

1

Corundum usually crystallizes in the form of striated fusiform (1, 4), barrel-shaped (5, 6), or tabular crystals. Aggregates are coarsely granular as well as compact. Interrupted growth has caused the transversal separation of corundum. It is not cleavable, has a conchoidal fracture, and on account of its hardness ranks next to diamond. It is of grey, brown-red, yellowish, or red colour, of feeble or pearly lustre, and is dull to opaque. Its precious varieties are usually translucent or transparent, and of high glass-like lustre. Leucosapphire is

98

2

a colourless, perfectly clear variety. The red ruby is coloured by a trace amount of chromium (2), the blue sapphire (3) by iron and titanium. Corundum is used as an abrasive. Its precious varieties, which are also produced synthetically, are mostly used for jewellery, in mechanical engineering, and in the construction of lasers.

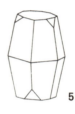

5

6

4

3

Haematite Fe$_2$O$_3$
Trigonal; H. 6.5; Sp.gr. 5.2—5.3

Haematite is a very common mineral. It has been mined since ancient times as an iron ore. The Babylonians and Egyptians used it for ornamental purposes. It occurs in nature in diverse forms, being mostly concentrated in ore deposits of magmatic, sedimentary, and metamorphic origin. Hydrothermal haematite is separated out on low-temperature veins and, having replaced limestones, it forms ore bodies on contact with igneous rocks. As an accessory mineral it crystallizes in the craters of some volcanoes or in lavas. Sedimentary haematite is of considerable economic importance. Vast haematite bodies in sedimentary rocks contain large amounts of workable ore, and for similar reasons metamorphic deposits of haematite are also of economic value. In these deposits the haematite originated by the metamorphosis of other iron ores such as limonite. Finely disseminated haematite causes the reddish coloring of many other minerals and rocks.

Haematite deposits of the above-mentioned types occur all over the world, the most important being in the USA, the USSR, and Sweden. Crystallized haematite is also found in many places, examples in Europe including Rio Marina, Elba; Traversella and Brosso, Italy; and Altenberg, GDR. Rose-shaped clusters of bladed crystals come from localities at Cavradi and Lercheltina, Switzerland, and are part of the so-called Alpine paragenesis.

5

4

Haematite forms many-faced, rhombohedral (5), or tabular (6) crystals. The foliated haematite named specularite (2) often forms the layered rock called itabirite. Haematite aggregates are granular, oolitic (4), or earthy (ochre), and radiate-fibrous with reniform to globular surfaces (kidney ore — 3). Crystallized haematite is of steel-grey to black colour, often with a beautiful iridescent tarnish (1) on the surface and of perfect metallic lustre; aggregates are

100

usually red and of dull lustre (shining ore). The mineral has a conchoidal fracture and a brown-red streak, and is brittle, with scaly separation. It is an important iron ore, red ochre being used as the raw material for making red paint and as a polishing agent; kidney ore has been valued for ornamental use. Similar minerals are magnetite, ilmenite, chromite, and also, sometimes, compact cinnabar.

Quartz and its varieties SiO$_2$

Hexagonal; H. 7; Sp.gr. 2.65; 46.7 per cent Si

Quartz is the commonest of all minerals, forming 12 per cent of the Earth's crust. It was the first mineral used by primeval man in the Stone Age. Flint, which is a cryptocrystalline siliceous mineral, has long remained an important trade article.

Quartz originates in many different ways. It is deposited from the magma as a part of igneous or volcanic rocks and pegmatites, is crystallized from solutions on hydrothermal ore veins, and is also formed during regional rock transformation (metamorphism). As a result of the weathering and decomposition of these rocks it also occurs in various sediments (e.g. quartzites and sandstones). It is also formed by deposition from solutions at normal temperatures, impregnates porous rocks, and sometimes results in the fossilization of organic remains (e.g. petrified wood). Common quartz originated in one of the above-mentioned ways and is very abundant. Rock crystal, smoky quartz, amethyst, and rose quartz are less commonly found. They are used in jewellery and in the manufacture of articles of aesthetic value.

Rock crystal forms in druse fissures in pegmatites, in ore veins, and in cracks in sediments. It is commonly found in the Alps in veins associated with minerals of the so-called Alpine paragenesis. In crystalline schists it often forms large 'rock crystal chambers'. One of them, found in 1735 by the Strahlers in Zinkenstock, Switzerland, yielded about 50,000 kg of rock crystal (some crystals weighed up to 500 kg).

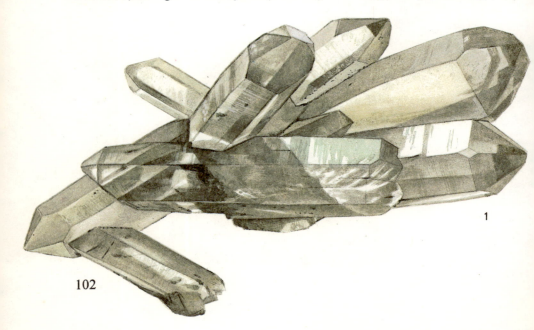

1

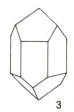

3

4

5

2

Quartz crystals and twins usually have the form of transversely striated prisms terminated by a rhombohedron (3, 4, 5). Aggregates are usually coarsely to finely granular to compact. Quartz is translucent, grey- to milk-white, dull, and of a glassy conchoidal to splintery fracture. Precious varieties of quartz have different optical properties. Rock crystal (1) is a vitreous, colourless variety of quartz. Because of its transparency and coolness when touched it was believed by ancient man to be a form of ice, as is testified by its name, derived from the Greek *krystallos,* ice. In recent times it has been of importance in glass-making,

optics, electronics, and radio technology. A considerable part of the world consumption is met by synthetic rock crystals. Smoky quartz (2) and the black variety called morion owe their colour to the 'colour centres' formed in the imperfect crystalline lattice, especially after irradiation by gamma rays.

103

Quartz and its varieties (continued)

Smaller crystals of rock crystal are found in Dauphiné, France, and still finer ones in the cavities of the Carrara marble in Italy. The so-called Marmarosh diamonds are found in alluvial deposits in Transcarpathian Ukraine, USSR. Abnormally large crystals of rock crystal are also found in Hot Springs, Arkansas, and Middleville, New York, USA, but the largest crystals of all have been found in Madagascar (some of these are several metres in circumference).

Smoky quartz is a transparent mineral of smoky brown colour, and is found associated with rock crystal in pegmatites, ore veins, and minerals of the Alpine paragenesis. Extraordinarily large crystals were found in 1868 in the Uri canton and near St Gotthard, Switzerland.

Amethyst forms on hydrothermal veins of many deposits. Associated with agate, however, it is found in amygdaloidal cavities in melaphyres. The most important deposits are in Idar-Oberstein, FRG; Kozákov, Czechoslovakia; Uruguay; and south Brazil. Around the turn of the century, a geode of $10 \times 5 \times 3$ m was found in Sierra do Mar, Brazil; it yielded 70,000 kg of druse crystals of the best quality.

Rose quartz is the mineral of some pegmatites. It forms coarsely granular aggregates; wonderful druses were found in 1971 in Arssuahy-Jeguitinhonha, Minas Gerais, Brazil.

The violet colour of amethyst (3) is due to the admixture of iron, in which Fe^{2+} oxidizes into Fe^{3+} under the effects of radiation. Similar physico-chemical disturbances in the crystalline lattice of rose quartz (2) are the cause of its colour.

2

4

Ferruginous quartz (Eisenkiesel) (4) is
coloured red by microscopic inclusions
of haematite, and yellow or brown by
inclusions of limonite. It most often
occurs in deposits of iron ores such as
those at Iserlohn, FRG; Ovieda, Spain;
and in central Bohemia. Another
interesting variety is the sceptre-shaped
quartz (1). It forms thin prismatic crystals
with a larger sceptre-shaped top. This
kind of quartz is found in Banská
Štiavnica, Czechoslovakia.

3

1

105

Chalcedony and its varieties SiO_2
Cryptocrystalline; H. 6.5—7; Sp.gr. 2.59—2.64

For thousands of years chalcedony has been one of the most popular minerals, used for decorative objects and amulets. The mineral was named after Chalkedon near Istambul, Turkey, where it was mined in ancient times. It is one of the subvarieties of silicon dioxide and is a cryptocrystalline mixture of parallel and radiate fibres of quartz, cristobalite, and other minerals with amorphous opal. It originates from low-temperature solutions near the Earth's surface, or by the decomposition of rock-forming silicates. The fibrous inner structure of chalcedony and its variable compactness mean that im more porous structures water and other solutions may be absorbed. Due to this infiltration and due to microscopic admixtures of some other minerals the precious varicoloured varieties originated. Examples include jasper, cornelian, and chrysoprase, but the most valuable is agate. Enhydros is a translucent nodule of chalcedony-agate; the hollow core remains partly filled with the original parent solution. It occurs in Uruguay and Canada. Jasper, which is the most common variety of chalcedony, is found, associated with cornelian, especially in areas of volcanic melaphyres and porphyres. Chrysoprase originates in the serpentinization of dunites. The most important deposit is in Ząbkowice in Poland.

Chalcedony also forms pseudomorphs after fluorite (e.g. in Trestia, Romania) and calcite, and produces petrified wood. On account of its hardness and power of resistance, it serves as excellent raw material in the manufacture of mortars, journals in bearings, and polishing powder.

3

1

Chalcedony is compact (1), of splintery fracture, translucent, of glassy or greasy lustre, and always rather dull. It has a reniform surface and sometimes forms semi-globular, botryoidal or ventricular aggregates, and slender stalactites. It is grey or grey-blue, sometimes yellowish, and concentrically shelled. The colourless variety is rare. Cornelian (2) is flesh-red in colour. The typical green colour of chrysoprase (3) is due to nickel liberated from olivines decomposed in the course of serpentinization. Jasper is a mixture of quartz and spherulitic chalcedony, its colouring depending upon the disseminated minerals. Haematite colours it red, ochre-limonite yellow, and inclusions of chlorite, actinolite, or serpentine green. The above-mentioned varieties of jasper are cut for ornamental purposes.

2

Chalcedony and its varieties (continued)

The Egyptians and Sumerians considered agate the most valuable ornamental stone, and its beauty has been the object of admiration up to present times. It was named after its first known locality along the River Achates (today Dirillo) in southern Sicily. Agates are composed of parallel bands of quartz, chalcedony and opal. Their typical rounded nodules have a finely banded texture of differently coloured layers, and are usually hollow. The walls of such cavities are often covered with radially grouped crystals of rock crystal and amethyst, their tops pointing to the centre of the cavity (a geode).

Agates originate in effusive rocks, filling vesicles left behind after the escape of gases, or filling veins by the gel of silicic oxide mixed with water. Penetrating solutions, enriched by iron and manganese, precipitate these dissolved substances and form varicoloured concentric bands. Agates usually display channels through which the solutions penetrated their interiors. The so-called moss agate contains dendritic forms of iron, manganese oxides, or chlorite.

Agate occurs in amygdaloidal melaphyres, and in similar volcanites. The most important deposits are in Idar-Oberstein, FRG. Agate was mined here as early as Ancient Roman times; nowadays the deposit is depleted and consequently of no commercial importance. Bohemian deposits in Kozákov and Nová Paka are also of no great importance any more. The most important producer of agates is Brazil; smaller deposits are found in Iceland, the Urals, USSR, and China.

3

Agate is a compact variety of chalcedony with fine concentric banding, or irregular variegated patterns, or parallel or wavy bands. The variegation of bands is caused by chemical admixtures and microinclusions of haematite, limonite and other minerals. Natural agates range from white, grey-blue, red (1, 2), yellowish, brownish to black in colour. Some bands are more porous, thus enabling different solutions to seep through them. On account of this property some agates, especially the Brazilian ones, are coloured artificially by aniline colours or other mineral pigments. The white-black-banded onyx originates from agate soaked in honey or sugar solution. When heated or immersed in sulphuric acid, the honey or sugar in more porous bands turns black. From the various kinds of agate, mention should be made of the speckled, ruin (3) and coral agates.

Common opal $SiO_2 . nH_2O$
Amorphous; H. 5.5—6.5; Sp.gr. 2.1—2.2

Opal is a typical amorphous mineral. It originates by the hardening of colloidal aqueous silicon dioxide into a glassy amorphous mass with an irregular arrangement of particles in its space lattice. Consequently, opal does not form crystals or aggregates but only compact masses with globular or reniform surfaces. Often, however, it changes due to recrystallization into chalcedony or quartz. The water content varies from 1 to 21 per cent, since the opal loses part of its water content in dry air and in damp air regains its moisture content again. Due to the disarranged space lattice it is often tinted by admixtures and various impurities. In this way many new varieties originate.

Opal forms in many different ways in nature. It is the product of hydrothermal decomposition of rock-forming silicates, especially in Tertiary igneous rocks such as andesites. It originates during the serpentinization of dark magmatites (e.g. dunites) and is usually associated with compact magnesite. As a product of the final stages of volcanic activity it often forms in cavities of young volcanic effusive rocks, or originates by deposition from hot springs and forms nodules and veins in sediments. It often impregnates organic remains and acts as petrification medium producing petrified wood.

Opal is a compact mineral of conchoidal fracture and occurs in many varieties in nature. Hyalite (1), from the Greek word *hyalos,* meaning glass, is an attractive mineral forming pellucid globular

1

2

3

aggregates, reniform or botryoidal
concretions, or dripstones. The mineral is
found in Valeč, Czechoslovakia; Cerro
del Tepozan, Mexico, and in Japan. Some
hyalites, for instance those from North
Carolina, USA, are luminescent, with
a green sheen in ultraviolet light.
Common opal is coloured by different
admixtures achieving rich shades of
milk-white, yellow, brown-red, green, and
grey-black. It is usually translucent or
opaque, and of glassy, waxy, or pearly
lustre. Moss opal (2) contains black
dendrites of manganese hydroxides;
wood opal (3) reveals the structure of
plant tissues. Many varieties of opal are
used as ornamental stones.

Precious opal

Precious opal has occupied an important position among precious stones for thousands of years, as proved by archaeological discoveries dating approximately from 500 B.C. On account of its remarkable colours it is a highly prized variety of opal. The play of colours shows even on the smallest pieces of the mineral. The physical cause of this phenomenon has long remained unexplained. It has been considered to be a result of the interference of light rays refracted by fine platelets of disseminated calcite and from water present in microscopic vesicles of the amorphous opal. The electron microscope, however, has revealed that the internal structure of precious opals is by no means chaotically arranged but forms a regular lattice similar to crystalline substances. It is composed of large spherical particles of identical diameter containing amorphous silicon dioxide, which are symmetrically grouped in solid siliceous gel with a variable content of water. The play of colours is due to the reflection of light by finely dispersed inter-particle vesicles filled with water. This reticulated structure of net-like cross-meshes of a diameter corresponding to the wavelength of light, scans the rays into spectrum colours, not into mixed interference colours, as was previously believed.

Precious opal is not very common. The most beautiful opals were those found in Dubník, Slovakia. However, then the fame of opals from Dubník was overshadowed by the discovery of opals in Zimapan, Mexico, where the fire opal is also found. The most important deposits of precious opals, however, are in White Cliffs, Barraco River, Bulla Creek, Australia, where the most prized variety — the black opal — is also found.

Precious opal (1) varieties display an excellent play of pastel as well as deep green, green-blue, blue-violet, yellow, and red colours. This is also due to the water content absorbed in the opal. Some opals lose part of their water content in dry air and consequently lose the play of colours, but immersed in water or in damp air they may regain their moisture content. Large specimens of precious opals are rare. One of them, called the Troy Fire (194 carats), is exhibited in

1

Paris, and the Harlequin from Dubník,
Czechoslovakia, (approximately 600 g) is
kept in the Natural History Museum in
Vienna. Fire opal (2) is translucent and of
a deep hyacinth-red to gold-brown
colour. Its opalescence can be increased
by polishing.

2

Rutile TiO$_2$
Tetragonal; H. 6—6.5; Sp. gr. 4.2—4.3; up to 61 per cent Ti

Rutile is the most common of the three modifications of titanium dioxide. It got its name in 1820 from the Latin *rutilus,* meaning yellow-red, after the colour of its lighter-coloured varieties. It is the most widespread titanium mineral. Hair-like crystals of rutile are often disseminated in sediments, larger grains and crystals form in quartz veins and quartz lenticles in gneisses. Famous locations where the mineral is found are near Soběslav, Czechoslovakia, and near Magnet Cove, USA. After the weathering of parent rocks rutile may be concentrated in alluvial deposits. It is mined from apatite veins near Kragerö, Norway, from pegmatites in Roseland, Virginia, USA, and in the Urals, USSR. Beautiful rutile crystals are found in veins associated with minerals of the so-called Alpine paragenesis, where it occurs in the form of long needles embayed in rock crystal (sagenite). Important occurrences are Modriach and Pfitsch, Austria; Binnental and St Gotthard, Switzerland; and Limoges, France.

Anatase TiO$_2$
Tetragonal; H. 5.5—6; Sp.gr. 3.8—3,9

Anatase occurs only in tiny, perfect crystals. It is mostly found in cavities and cracks in crystalline schists together with other minerals of the so-called Alpine paragenesis, particularly rock crystal and adularia. The most important localities are Bourg d'Oisans, France; Praděd, Czechoslovakia; Hof, GDR; Minas Gerais, Brazil; and Colorado, USA. It sometimes forms a microscopic admixture in some igneous rocks and sediments.

Brookite TiO$_2$
Orthorhombic; H. 5.5—6; Sp.gr. 3.9—4.2

The less common brookite occurs only in crystals and accompanies anatase in almost all its occurrences. It is an unstable mineral and often transforms into rutile.

Rutile crystals are columnar, longitudinally striated, and often form heart-shaped, knee-shaped (1), or cyclic twins (4). Hair-like needles form reticulated aggregates (sagenite). Rutile is cleavable, of irregular fracture in aggregates, brittle, translucent or opaque, and of submetallic or adamantine lustre.

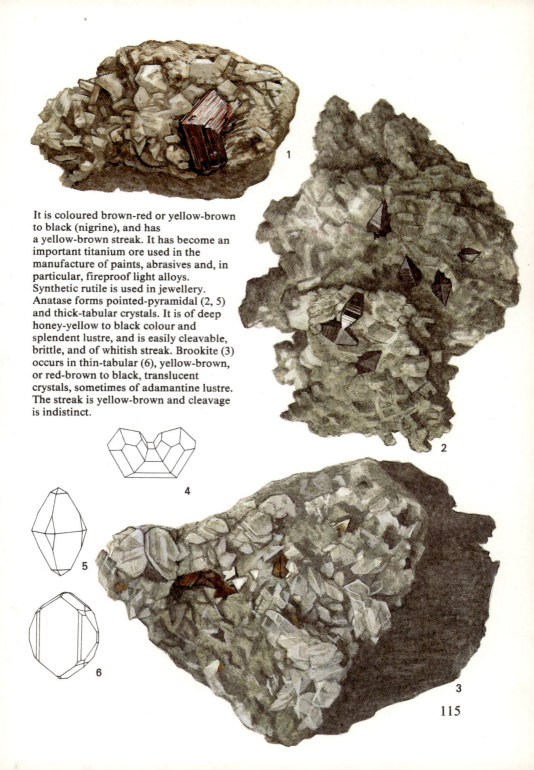

It is coloured brown-red or yellow-brown to black (nigrine), and has a yellow-brown streak. It has become an important titanium ore used in the manufacture of paints, abrasives and, in particular, fireproof light alloys. Synthetic rutile is used in jewellery. Anatase forms pointed-pyramidal (2, 5) and thick-tabular crystals. It is of deep honey-yellow to black colour and splendent lustre, and is easily cleavable, brittle, and of whitish streak. Brookite (3) occurs in thin-tabular (6), yellow-brown, or red-brown to black, translucent crystals, sometimes of adamantine lustre. The streak is yellow-brown and cleavage is indistinct.

Cassiterite SnO_2
Tetragonal; H. 6—7; Sp.gr. 6.8—7.1; 78 per cent Sn

Cassiterite was one of the first minerals known to man and has been used since ancient times. The Bronze Age findings show that some 8,000 years ago man learned to melt tin from cassiterite, and with the admixture of copper produced a very useful alloy.

Cassiterite is the richest and most important tin ore. It occurs in primary deposits, and secondarily in alluvial placer deposits. Primary cassiterite originates in the mineralization of cooling granitic magma. As an accessory mineral it occurs disseminated in granite and in larger amounts it is concentrated in granite pegmatites; workable deposits are found especially in the greisens. These rocks were formed by the decomposition of granites by hot, gas-charged solutions, and contain quartz, zinnwaldite, topaz, apatite, fluorite, molybdenite, cassiterite, wolframite, and sometimes also scheelite and arsenopyrite. The above-mentioned mineral association may also occur in rocks overlying granites or in quartz veins. Some of these high-temperature veins also contain chalcopyrite, stannine, pyrite, magnetite, sphalerite, and galena, and in Bolivia they also contain silver ores.

Primary deposits of cassiterite occur in many areas with granitic intrusions, e.g. in the Bohemian and Saxon Ore Mountains; in Bretagne, France; and in Cornwall, Great Britain. Rich deposits are found in Africa, Australia, and South America. Almost a half of the world production comes from secondary deposits in Indonesia and Malaysia.

3

Cassiterite usually forms short-columnar (4) crystals and heart-shaped twins (1, 2, 5) and compact to coarsely granular aggregates. Fibrous masses are called 'wood tin' (Holzzinn). Cassiterite is black-brown in colour, of strong submetallic lustre, brownish to whitish streak, and indistinct cleavage, and is fragile and translucent to opaque. Grey, greenish, yellow, and hyacinth-red shades are rare. Synthetic cassiterite is colourless. Cassiterite from alluvial placers occurs in the form of irregular pebbles (3). Cassiterite is the most important tin ore. Tin was previously used in the production of cooking vessels; nowadays it is mostly needed for the manufacture of tin foils and tins for the canning industry, and for the manufacture of special alloys. Similar minerals are rutile, sphalerite, garnet, vesuvianite, and zircon.

1

4

5

2

Pyrolusite MnO_2

Tetragonal; H. up to 6 (varying); Sp.gr. about 5 (varying); about $1-2$ per cent H_2O

Pyrolusite belongs to a complex group of manganese oxides and hydroxides, which can only be distinguished with difficulty. Related minerals include polianite, braunite, manganomelane-psilomelane, and wad. All of them originate secondarily by weathering of the primary rhodochrosite and rhodonite or minerals with an admixture of manganese (e.g. siderite, ankerite, and dolomite). They are of sedimentary or residual origin, and are found on the bottom of oceans, forming nodules which contain workable amounts of various metals.

Pyrolusite and psilomelane are the most important manganese ores. They are found in the oxidation zones of primary manganese ores, e.g. in the siderite veins in Siegerland, FRG; in Ilmenau and Ilfeld, GDR; and in the quartz veins with haematite near Horní Blatná, Czechoslovakia. Large accumulations of pyrolusite and psilomelane are of sedimentary origin (e.g. the deposit at Cziatura, USSR). Similar deposits are in Batesville, Arkansas, USA; Ouro Preto, Brazil; Kodur, India; and in Africa. Both minerals were used by the Ancient Romans to colour glass. At present they are an important raw material for the production of ferro-manganese for the metallurgical, chemical, and electrochemical industries.

2

4

Pyrolusite rarely forms columnar crystals (4); more often it occurs in the form of porous radial and fibrous aggregates (1), and pseudomorphs after magnetite and polianite. It is grey-black, of submetallic or dull lustre, brittle, perfectly cleavable, and of black streak. Psilomelane (manganomelane) is amorphous, forms botryoidal bunches, reniform crusts (2), dripstones, and earthy coatings. It is black, of brown-black streak, and dull to waxy lustre. Due to the high porosity of aggregates different ions are concentrated in psilomelane; consequently, some psilomelanes contain important amounts of barium, copper, cobalt, nickel, vanadium and other elements. Earthy or powdery wad often forms dendritic crusts (3) in cracks.

1

3

Wolframite (Fe,Mn) WO$_4$

Monoclinic; H. 4.5—5.5; Sp.gr. 7—7.5

Wolframite has a varying content of iron and manganese, and is the most important wolfram (tungsten) ore. For many centuries miners considered it gangue material associated with cassiterite, and an unwelcome admixture in melting tin ores. It is said to have been named by Saxon miners who noticed that in the process of melting, it 'swallowed' (like a wolf) the tin-bearing foam ('Rahm' in German).

Wolframite originates in the vicinity of granite intrusions; it was separated out from hot solutions in pegmatites, greisens, and quartz veins, as a constituent of the tin-molybdenum-wolfram mineral association. Typical associated minerals are cassiterite and molybdenite. The mineralization process was dealt with in more detail in the pages describing these other minerals.

Wolframite deposits are found in areas with massifs of intrusive granites. In Europe the most important deposits are in Altenberg, Annaberg, and Schneeberg, Saxony, GDR; Cínovec, Krupka, Horní Slavkov, in the Bohemian Ore Mountains, Czechoslovakia; Cornwall, Great Britain; Beira Baixa, Panasqueira, Portugal; and Sierra Almagrera, Spain. In Neudorf (FRG) wolframite occurs in galena-bearing veins, and it has also been mined near Limoges, France, and from quartz veins containing beryl in Andunczilon, USSR. Large deposits are found in Kiangsi, China: Boulder County, USA; Bolivia; Brazil; Sierra de Cordoba, Argentina; Africa and Australia.

2

1

3

4

Wolframite crystallizes in short-columnar and thick-tabular (1, 3), twinned (4), striated and sometimes very large crystals. In Panasqueira crystals more than 70 cm long were found. It forms granular, spathic, and long lath-like (2) aggregates. It is coloured black with a brownish tinge, has a brown-black streak and metallic lustre, is opaque, brittle, and perfectly cleavable, and sometimes has shelled separation. It is the source of wolfram, which is used in many industries, especially in the manufacture of hard-tool steels, filaments for electric light bulbs, luminophores for X-ray tubes, dyes for the textile industry, and non-combustible fabric impregnations.

Uraninite empirically U₃O₈ (mixture of oxides)
Cubic; H. 4—6; Sp.gr. 9—10

Uraninite has become one of the most valuable strategic raw materials although it was formerly considered quite unimportant. Miners in Jáchymov, Czechoslovakia, and in the Saxon Ore Mountains believed it to be a useless constituent in ore veins and a sign indicating gangue material. Their experience showed that the occurrence of uraninite in a vein usually meant the disappearance of rich silver ores. The pitch-black mineral was believed to bring bad luck to miners, and they called it pitchblende, and threw it away as unimportant. It was used for the first time in 1850 in Jáchymov for the manufacture of luminous paints and stained glass. After the discovery of radioactivity, and, later on, of the rare elements radium and polonium in the Jáchymov ore, uraninite became an important raw material in their recovery. Its value increased further with the discovery of the fission of uranium, a process used in the nuclear industry.

Uraninite occurs in two types of deposits. The most frequent yet less productive are pegmatites, in which it is often found crystallized, e.g. in Moos, Norway; Branchville, Connecticut, USA; Wilberforce, Canada; and Uluguru, Namibia. Much more important sources are hydrothermal ore veins where it often forms a compact vein filling, and it is also found as a constituent of the nickel-cobalt-bismuth-silver-uranium formation, or associated with copper ores. The most important deposits include Jáchymov, Czechoslovakia; Great Bear Lake, Canada; and Shinkolobwe, Shaba, Zaire.

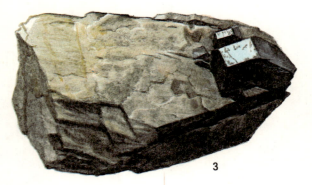

3

4

Crystallized uraninite with an increased content of thorium is called bröggerite (3). It most often forms cubes, octahedra, or their combinations (4). Uraninite crystals are found in ore veins (e.g. in Shinkolobwe, Zaire). Compact uraninite (2) occurs in the form of reniform aggregates (1) of conchoidal fracture and brown-black streak. The mineral is opaque, black, of resinous, submetallic or greasy lustre, and is fragile. In small amounts it is also found disseminated in intercalations of coal, e.g. in some pyrite slates in Sweden.

Uraninite weathers easily, and the first product of the decomposition is usually the amorphous, earthy uranium black. In damp air uraninite quickly becomes coated with varicoloured secondary minerals.

2

1

Bauxite (mixture of aluminium minerals)
H. 1—3; Sp.gr. 2.4—2.5

Bauxite, a sedimentary rock, is in fact a group of related aluminium oxides and hydroxides, such as the amorphous alumogel, monoclinic gibbsite (hydrargillite) and the orthorhombic diaspore with boehmite. In isolation these minerals occur only in small amounts and are of no economic importance; in contrast, bauxite has become the most important raw material in the production of aluminium.

Bauxite originates secondarily by the decomposition of rock-forming minerals rich in aluminium. It forms under special conditions as a result of the destruction of the bond between aluminium and silicon in the original aluminosilicate. The released silicon is separated out as opal or quartz and, like other admixtures of the parent rock, such as TiO_2 (up to 3 per cent), and especially haematite, it becomes a part of bauxite.

Bauxite was named after the locality where it was found for the first time, Les Baux in France. Large deposits are also found in Italy, FRG, Yugoslavia, Romania, Hungary, the USA, Guyana, India, Jamaica, Australia, and elsewhere.

The soil accompanying bauxite in karst areas is called *terra rossa*. Geological and climatic conditions favourable for the origin of bauxite occur in tropical regions, with high temperatures and alternating dry and rainy periods. Large deposits also occur in argillaceous limestones in karst areas; workable isolated accumulations are the product of decomposition of nepheline rocks under the effects of sulphuric acid (e.g. at Pulaski, Arkansas, USA).

Bauxite forms earthy or compact masses of conchoidal fracture, often preserving the structure of the rocks from which it originated. Pure bauxite usually has an oolitic (pea-like) or pisolitic texture, or occurs in irregular nodules. Its colour varies because of admixtures of other substances. It is mostly of light shades, whitish, yellowish, and brown (1), but may also be red or even black. **Laterite** is a bauxite variety containing up to 30 per cent Fe_2O_3. Its deposits in tropical regions cover large areas of land. It is a brick-red to brown-red earthy soil (2).

Bauxite is the most important raw material providing aluminium for the electronics, mechanical engineering, aircraft and nuclear industries. It is also used in the production of abrasives, fireproof materials and cement.

1

2

Limonite FeO(OH) . nH$_2$O
H. 5—5.5; Sp.gr. 2.7—4.3; about 40—63 per cent Fe

Limonite is the common name of a whole group of amorphous as well as crystalline iron hydroxides with a varying water content. It forms many varieties which differ widely in their appearance, e.g. brown and yellow ochres, stilpnosiderite, goethite, velvet ore, lepidocrocite, and others. Roentgen analysis, however, proves the existence of only three separate mineral varieties, i.e. goethite (orthorhombic α-Fe O(OH)), lepidocrocite (orthorhombic γ-FeO(OH)) and an amorphous gel variety.

Limonite is common in occurrence, and is usually the decomposition product of iron minerals. Usually it occurs close to the Earth's surface. On outcrops of ore deposits it forms the 'iron hat' (also known as iron cap, ironstone, or gossan) containing secondary minerals of non-ferrous metals, and sometimes also native gold and silver. As a result of hydrothermal mineralization limonite sometimes crystallized in ore veins in the form of primary goethite. It also separates out from peat-moor water ('turf ore') and is deposited on the floors of some lakes. Then it is called lake ore (or bog iron ore or bean ore). These deposits are formed by the action of bacteria. Large deposits of sedimentary oolitic limonite occur in Lotharingia and are called 'minettes'. From a great number of commercially important deposits mention should be made, at least, of Rožňava, Czechoslovakia; Priedor, Yugoslavia; Spain; Siegen, FRG; Sweden; Finland; and Canada.

4

Goethite (needle ironstone, acicular iron ore) forms acicular crystals (5) and fan-like, fibrous, reniform aggregates (1); velvet ore coats other minerals with rusty-coloured hairy crusts (2). The typical form of the less common lepidocrocite is rose-shaped platelets; it also occurs in powdered or reniform form. Both minerals have perfect cleavage, yellowish to red-brown colour, metallic lustre, are translucent in thin slabs, and have a brown to brown-yellow streak. The black, coarsely reniform stilpnosiderite usually has resinous lustre.

Compact limonite also forms botryoidal
masses and thin stalactites (3). The earthy
bog iron ore (4) is porous; powdery
ochres are non-cohesive and loose. Man
learned how to work limonite in ancient
times, and in the Middle Ages it
remained a source of iron for smaller
hammer-smith manufacturers. Nowadays
it is only economically workable if it
occurs in very large amounts.

5

3

2

1

127

Smithsonite $ZnCO_3$
Trigonal; H. 5; Sp.gr. 4.3—4.5; up to 52 per cent Zn

Smithsonite is an attractive, varicoloured mineral and an important zinc ore. It has been used in the production of brass (which is an alloy of copper and zinc) since the Middle Ages, long before metallic zinc was recovered, and before smithsonite was distinguished as a separate mineral at the end of the 18th century. Smithsonite was named in honour of the English chemist J. Smithson.

It originates as a secondary mineral by the alteration of sphalerite and other primary zinc ores. If the circulating subsurface waters contain dissolved zinc sulphate, zinc separates out in the form of a carbonate — smithsonite — in places where they penetrate limestones or dolomites. By the replacement of these carbonates large deposits of metasomatic smithsonite are formed. The majority of large deposits, in fact, occur in carbonaceous rocks containing primary polymetallic ores. Associated minerals — in a mixture called galmei by miners — are usually hydrozincite, hemimorphite, willemite, and some secondary minerals of lead. Admixtures of trace amounts of cadmium in sphalerite form the secondary yellow greenockite (CdS). Some smithsonites containing up to 3 per cent cadmium are an important cadmium ore. Large deposits are found in Lavrion, Greece; Bleiberg, Austria; Olkusz, Poland; Iglesias, Sardinia; Vieille Montagne, Belgium; Tsumeb, Namibia; Broken Hill, Australia; and in the USA.

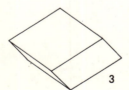

3

Crystallized smithsonite usually occurs in the form of small, imperfectly developed rhombohedral crystals (3), often rounded, with coarse faces, and perfectly cleavable. It is also commonly found as compact reniform (1) and botryoidal aggregates with a triangular or orthorhombic pattern on the surface. It also forms fibrous crusts, sometimes of agate-like texture (2), and even earthy forms occur. It has an uneven fracture and vitreous to pearly lustre, is fragile, and is translucent to clouded. Pure smithsonite is colourless, often whitish

1

and grey, but is coloured yellow, orange, brown, green, blue, or black-grey by admixtures. Similar minerals include some cellular calcites, hemimorphite with the characteristic oblong pattern on the surface of its reniform aggregates, and phosphorite.

2

Siderite $FeCO_3$
Trigonal; H. 4—4.5; Sp.gr. 3.7—3.9; up to 48 per cent Fe

Siderite is an iron ore of high quality, economically workable also because of a low content of various useful admixtures such as sulphides, which are, however, undesirable in iron-metallurgy. It is very common, and in some places forms large accumulations of commercial importance. The best example of a siderite deposit is the hill called Erzberg near Eisenerz, Austria, which has been worked since the 8th century. It is entirely composed of siderite, and the mineral is exploited here in open pits.

Siderite originates in magmatic and sedimentation processes. It is deposited from hydrothermal solutions in ore veins of different metals, or it forms large bodies of relatively pure ore by the metasomatic replacement of limestones and dolomites. It is found in ore veins in Siegen, FRG; Neudorf, GDR; Příbram, Czechoslovakia; and Cornwall, Great Britain. Metasomatic deposits are found in Erzberg, Hüttenberg, Austria; Schmalkalden, GDR; and Bilbao, Spain. Sedimentary siderite occurs in Schmiedefeld, FRG; and Masabi Range, Minnesota, USA. When mixed with clay, it forms loaf-shaped concretions, nearly half a metre long, called clay ironstone. They are usually found in coal basins; examples are known in Great Britain, the FRG, and Czechoslovakia.

2

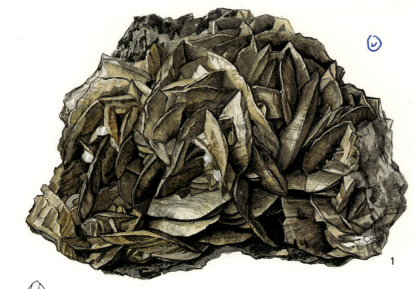

1

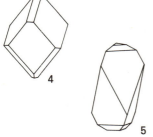

4

5

3

Siderite often crystallizes in the form of rounded rhombohedra (1, 4, 5); most often, however, it forms granular or compact aggregates, and sometimes is even oolitic or earthy. It is yellowish or brown in colour; if containing the coal admixture called blackband it can also be black. It has a good cleavage and is fragile with glassy or pearly lustre, and is opaque or sometimes translucent. In compact masses it has a conchoidal fracture. It weathers easily into limonite. Loaf-shaped concretions of clay ironstone (2) sometimes contain crystals of millerite, whewellite, and other accessory minerals.

Ankerite is a carbonate of calcium with an admixture of iron and manganese, and is associated with siderite in most deposits. Rhombohedral ankerite crystals (3) are grey-white, and on weathered surfaces are covered with rusty-brown coatings. Minerals similar to siderite are dark calcites and dolomites, and sometimes sphalerite may also be confused with it.

Rhodochrosite (dialogite) $MnCO_3$
Trigonal; H. 4; Sp.gr. 3.3—3.6; up to 48 per cent Mn

Rhodochrosite occurs in ore veins as well as in sedimentary or meta-morphic rocks. It was separated out from hydrothermal solutions mostly in the course of Tertiary volcanic activity, and is associated with ores of gold, silver, and polymetals in deposits such as Săcărâmb and Cavnic, Romania; Banská Štiavnica, Czechoslovakia; Silverton and Cripple Creek, Colorado, and Butte, Montana, USA; Freiberg, GDR; and Usinsk, USSR. Sedimentary rhodochrosite is usually found in deposits of pyrite slates, e.g. in Chvaletice, Czechoslovakia; and Huelva Province in the Pyrenees, Spain. It is also found in the metamorphosed deposit in Franklin Furnace, New Jersey, USA. It is mined as an important manganese ore. In South America it is used as an ornamental stone.

Rhombohedral crystals of rhodochrosite (1, 3) are usually imperfect and of drusy surface. Granular, compact, or fibrous aggregates form reniform crusts and botryoidal masses. Rhodochrosite (from the Greek words *rhodon*, pink, and *chrosis*, colour) is pink, red, or brownish, perfectly cleavable, of vitreous lustre, and translucent. It becomes coated by black manganese hydroxides on weathered surfaces.

1

2

Magnesite MgCO₃

Magnesite $MgCO_3$
Trigonal; H. 4—4.5; Sp.gr. 3; up to 47 per cent Mg

The economic importance of magnesite was discovered only relatively recently with the development of many different branches of industry. Its numerous deposits belong to two different genetic types, the products of which also differ in appearance. Crystalline magnesite is found rarely and in small amounts in eruptive rocks and pegmatites; workable accumulations, however, are found in limestones and dolomites, due to the metasomatic effects of hydrothermal solutions. The largest deposits of this type in Europe are in Veitsch, Leoben, Austria; and Jelšava and Ochtiná, Czechoslovakia. It is also found in Greece, Italy, and the USA. Compact chalk magnesite originated by the serpentinization of basic rocks. In serpentines, it usually occurs associated with newly created opal and talc. Apart from the Euboea deposit in Greece, such deposits are of no great economic importance.

Crystallized magnesite (2) has the form of simple rhombohedra or many-faced combinations (4). Usually it occurs, however, in the form of crystalline, perfectly cleavable aggregates, or compact masses of conchoidal fracture or even as earthy deposits. It has a whitish, blue-grey, yellowish, or brown colour.

Compact nodules and veinlets are chalk-white. Magnesite is used in the production of fire-proof bricks, and in ceramics. It is the source of magnesium, used in photoflashes and in the manufacture of light, non-corrosive (stainless) alloys for aircraft, ships, and other vehicles.

1

2

Calcite CaCO₃

Calcite $CaCO_3$

Trigonal; H. 3; Sp.gr. 2.6—2.8

After quartz, calcite is the most common mineral found in the Earth's crust. Apart from the crystalline mineral, it also occurs in the form of massive or compact rocks (limestones), sometimes forming vast mountain ranges. Man has known and used limestone since ancient times, as shown by the temples, palaces and sculptures made of marble, the metamorphosed crystalline variety of limestone. Calcite originates in a great many ways. It occurs in magmatic, volcanic, sedimentary, and metamorphic roks, and also forms secondarily by the decomposition of minerals rich in calcium. According to its origin, it ranges in appearance from crystals to granular or earthy aggregates, porous crusts, and other bizarre forms. In eruptive rocks, primary calcite is included only in some nepheline syenites. Larger amounts of magmatogenic calcite are found in cavities in volcanic and other rocks, where the calcite mostly crystallized from hydrothermal solutions on various veins. Sedimentary limestone has been deposited in thick layers on sea or lake bottoms due to the activity of living organisms. In the course of recrystallization (e.g. in mountain-building processes or at the contact with eruptive rocks) these sediments transformed into crystalline limestone, called marble. Karst phenomena show that limestone is soluble in surface waters, and therefore reenters the long-lasting migration cycle.

2

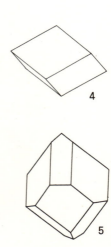

4

5

134

3

6

Calcite forms perfect crystals of various forms (1), and occurs in approximately 1,200 different crystallographic combinations. Its crystals are rhombohedral (4, 5), scalenohedral (2, 6), columnar (3), or acicular, tabular, or bladed.

1

Calcite (continued)

In contrast to limestone, crystallized calcite had been of no practical application for centuries. However, the discovery of many large limpid crystals in vesicles in the basalts near Helgustadire, Iceland, enabled the Danish physician Erasmus Bartolinus in 1669 to study for the first time the birefringence of light. A rhombohedron of limpid calcite is perfectly transparent and gives a marked double refraction (double-contoured picture) when one looks directly through it. Because of these properties Iceland spar has become an important raw material for the production of special optical apparatus, e.g. polarizing microscopes used in petrography. Deposits of calcite, limestone, and marble are found all over the world , and it is beyond the scope of this book to cite even the most important of them.

A special variety is the so-called tack-head spar formed of parallel twins of flat rhombohedra arranged so that the uppermost crystal is the largest (1). Granular, fibrous, oolitic, or compact aggregates form large masses, crusts, dripstones, and porous sinters. Calcite is perfectly cleavable, and has vitreous, pearly, or silky lustre. Pure calcite is clear or dull white, but impurities or inclusions produce different colour shades. The admixture of manganese, lead, etc., produces light blue, pink, and red luminescence in some calcites. Transparent crystals of Iceland spar (2) produce a marked double refraction. Calcite finds many different uses according to its purity. Clear, transparent crystals are used in the production of Nicol prisms for optical apparatus and lime and cement are produced from limestone. Various forms of calcite are used in glass-making, ceramics, the sugar industry, as a flux in iron smelting, and as an important raw material in the chemical industry.

2

1

Dolomite CaMg(CO₃)₂

$CaMg(CO_3)_2$

Trigonal; G. 3.5—4; Sp.gr. 2.8—2.9

Dolomite is a very common mineral and rock. It is very similar to calcite and was distinguished from it in the 18th century by the French geologist D. Dolomieu, after whom it was later named. Sometimes whole mountain ranges are composed of it. It originates by the deposition of shells of tiny sea animals in oceans, and particularly by the metasomatic replacement of non-cemented limestones affected by the action of sea water in the course of dolomitization. It occurs as a mineral in hydrothermal ore veins, and also crystallizes in serpentines, sediments, and talc schists. It is found in almost the same deposits as calcite. Dolomitic rocks are used as a cheaper alternative of magnesite in metallurgy, and in the production of special limes, mortars and cements.

Dolomite usually crystallizes in the form of rhombohedra (2), which are sometimes saddle-shaped (1). However, it also commonly forms granular or compact aggregates. In Binnental, Switzerland, the so-called sugar-dolomite has been found. It is perfectly cleavable and fragile, and its aggregates have conchoidal fracture. The colour of dolomite varies according to admixtures and inclusions. The purest dolomite is colourless, but the common forms are flesh-like, yellowish, or of brownish shades.

1

2

137

Aragonite CaCO₃
Orthorhombic; H. 3.5—4; Sp.gr. 2.95

Aragonite was classified as a separate mineral in 1790 and was named after its best-known deposit in Aragon, Spain. Miners knew it since the 15th century from Erzberg in Austria, but it was long mistaken for calcite, which is quite understandable because they both have the same chemical composition. Aragonite has many different varieties. Under special circumstances it may even change into calcite, which is the more stable calcium carbonate modification.

Aragonite originates by many different processes, but, unlike calcite, it is neither a primary constituent of rocks nor a typical mineral of ore veins. In exceptional circumstances clusters of crystals are found (e.g. in hydrothermal deposits in Špania Dolina, Czechoslovakia; and Alston Moor, Great Britain). In association with zeolites it formed rich druses in vesicles and cavities in volcanic rocks, particularly those formed towards the end of Tertiary time. The most important deposits are in Hořenec and Valeč, Czechoslovakia; Iceland; Nugsuak, Greenland; and many other areas of recent volcanic origin. It is mostly formed by deposition from hot springs, e.g. in Karlovy Vary, Czechoslovakia. Aragonite associated with calcite, gypsum, coelestine, or barytes is the product of complicated mineralization processes giving rise to stratiform sulphur deposits. Fine druses have been found in Girgenti and Agrigento, Sicily.

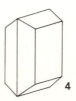

Aragonite crystals usually have the form of small coffins (2, 4), high pointed columns (5), or needles. Very often they form cyclic twins grown into six-sided prisms with their sides slightly concave in their central part and with blunt edges (1). Aragonite rarely forms granular or compact aggregates; most often they are acicular, and radiate to fibrous.

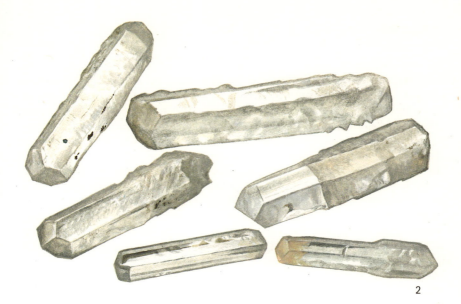

2

Coarse, parallel rope-like aggregates of aragonite (3) have no diagonal cleavage, a characteristic which distinguishes them from similar aggregates of calcite. They have only uneven fracture. Acicular aggregates sometimes show smooth separation surfaces along their length. Similar minerals, apart from some calcites, are barytes, strontianite, coelestine, and natrolite.

3

Much smaller amounts of aragonite crystals are found in the deposit of native sulphur in Machów, near Tarnobrzeg, Poland. The mineral occurs here also as isolated crystals and twins, often associated with gypsum and embedded in clays. Typical deposits of this type are in Molina, Aragon, Spain; and Bastennes, France.

Aragonite also originates secondarily by the decomposition of siderite under the effects of surface water in large deposits of iron ores, or in karst areas built of dolomitic limestones. In cavities in rocks and ores it occurs in the form of characteristic shrub-like or skeletic clusters. These were called *flos ferri* by miners in the Middle Ages. Large aggregates of 'iron flowers' are found especially in Erzberg and Hüttenberg, Austria, and in the aragonite cave at Hrádok, near Ochtiná, Czechoslovakia. Interesting varieties of aragonite may originate by deposition from hot springs, e.g. in Karlovy Vary, Czechoslovakia. Apart from the porous sinter and white balls found freely on the floors of cavities, large oolitic aggregates of peastone (pisolite, pea grit) have formed here. A large, several-metre-thick sheet of sprudelstein showing parallel white, cream, and brown-red bands covers the floor around the main hot spring. Aragonite has even been continuously deposited inside the water pipes draining mineral water from springs, and consequently the pipes must often be replaced by new ones.

3

2

An unusual form of non-crystallized aragonite is *flos ferri*, a shrubby growth of pure white aragonite (1). Other forms are the peastone composed of small pea-like spherical grains or regular balls (3) cemented together, and the so-called sprudelstein displaying a characteristic parallel banding of

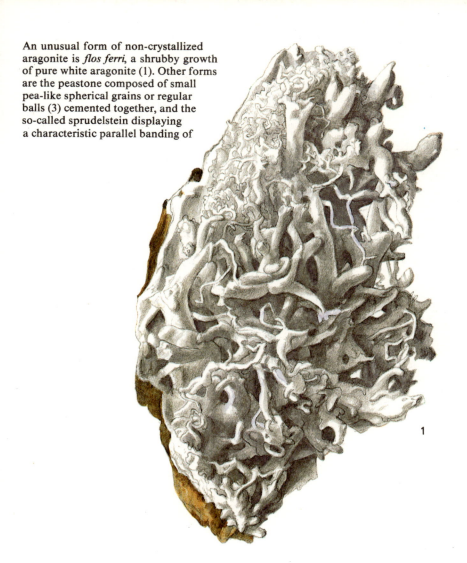

1

different shades (2). Aragonite film quickly forms on objects immersed in thermal waters. In this way 'stone roses' are produced from fresh flowers. The purest variety of aragonite is colourless to clear; it is, however, more often dull white, honey-yellow, and grey, but may also be pink and greenish. If it contains some dark impurities it is usually bluish or grey-black. It has vitreous lustre and may be transparent, translucent, or opaque.

Cerussite PbCO₃

Orthorhombic; H. 3—3.5; Sp.gr. 6.4—6.6; up to 77 per cent Pb

Cerussite derives its name from the Latin *cerussa*, meaning white, and was so named in 1845. It had been found by miners since Roman times near the surface of some deposits. It is exclusively a secondary mineral formed by the decomposition of galena and other primary lead ores. Workable deposits are found in limestones in particular. In small amounts cerussite also crystallizes in abandoned heaps of polymetallic deposits.

Cerussite, being found in places where polymetallic ores occur, is a widely distributed mineral. In mines it occurs only on exposures of ore bodies and in newly uncovered deposits. In old mining districts it is rather rare.

In Europe it has mainly been found in ore deposits in the GDR; Harz, Siegen and near Aachen, FRG; Příbram and Stříbro, Czechoslovakia; Olkusz and Bytom, Poland; Kirlibaba, Yugoslavia; Leadshill, Scotland; Iglesias, Sardinia; and Beresovsk and Nertschinsk, Transbaikal, USSR. Large deposits are found in Tunisia; Kabwe (formerly Broken Hill), Zambia; and in Tsumeb, Namibia, where elongated crystals tens of centimetres long are found. Large quantities of cerussite are also found in Arizona and Colorado, USA.

3

Cerussite is often found crystallized and occurs in about 90 different forms. Its crystals are usually tabular, or columnar (1, 3) to acicular, are fragile and without cleavage, and form typical contact or interpenetration twins. Crystal faces of bright adamantine lustre display fine striations. Granular, stellate, or compact aggregates have conchoidal fracture, greasy lustre, and form crusts or coatings sometimes having a reniform surface.

142

Cerussite may be colourless or clear, but also milk-white, grey, yellowish, or brownish (1). The mineral is translucent, rarely transparent, and has a white streak. The admixture of galena or graphite causes its grey-black (2) colouring. Similar minerals are anglesite, coelestine, barytes, and scheelite.

Azurite $2CuCO_3 . Cu(OH)_2$
Monoclinic; H. 3.5—4; Sp.gr. 3.7—3.9; up to 52 per cent Cu

Malachite $CuCO_3 . Cu(OH)_2$
Monoclinic; H. 3.5—4; Sp.gr. 3.9—4.1; up to 57 per cent Cu

Primary sulphur ores of copper have a typical metallic surface of an inconspicuous colour, whereas secondary minerals of copper (such as azurite and malachite) display bright colours especially of various blue and green shades. Both these minerals occur in almost all copper ore deposits. They usually formed by the action of solutions saturated with copper sulphate on calcite veins or limestone rocks. Malachite is more common than azurite, which usually forms first and then sometimes changes into malachite in a partial or complete pseudomorphosis. Together with other weathering products, both carbonates fill fissures and vesicles. Perfect crystals or thick crusts are found especially in cavities in limonite. Disseminated malachite and azurite impregnate large masses of sandstones and schists, and consequently build important stratiform deposits.

Both minerals are quite common, but in ores with an increased amount of primary tetrahedrite, azurite is more frequent. Perfect azurite crystals come from Chessy, France; Moldova, Romania; Lavrion, Greece; Tsumeb, Namibia; Bisbee, Arizona, USA; and Bura-Bura Mine and Broken Hill, Australia. In these and many other deposits malachite crystals may also be found. Malachite forms thick reniform encrustations and layers in the deposits in the Urals, USSR. In about 1820, a massive block measuring roughly 3×6 m and weighing about 250 tonnes was found in Nizhni Tagil, USSR.

5

Many-faced crystals of azurite usually have a rhombic appearance (1, 2), form plates, and often parallel twins (3). Aggregates are granular, radiate, and form kidney-shaped or botryoidal encrustations, pulverulent coatings, and pseudomorphs (e.g. after cuprite). Azurite was named after its deep azure-blue colour (from the Persian *lazhward*, blue). It is translucent or opaque and of glassy lustre, and has a blue streak and uneven fracture.

Malachite crystals are rare and form acicular bunches (4) or radiating clusters. Granular or fibrous aggregates often have a reniform surface with agate-like texture (5). Earthy and powdery aggregates are frequent. Malachite has a deep green colour, a lighter streak, and vitreous or silky lustre, and is cleavable and fragile. Both minerals are important ores of copper. Malachite is often used in jewellery and for ornamental and decorative articles.

Gypsum $CaSO_4 . 2H_2O$ ✓
Monoclinic; H. 1.5—2; Sp.gr. 2.3—2.4

Anhydrite $CaSO_4$
Orthorhombic; H. 3—4; Sp.gr. 2.8—3

Gypsum is the most common mineral of the sulphate group. It forms thick strata which are large enough to be classed as rocks. Most gypsum has been precipitated, together with anhydrite, from evaporating sea and salt-lake waters as a chemical sediment. It may also form in volcanic areas where sulphurous gases (sulphur oxides) have escaped; there the sulphuric acid destructively attacked the neighbouring carbonate rocks, especially limestones, transforming them into layers of gypsum. Gypsum may also be the product of oxidation of pyrite or marcasite. Occurrences of this type are very common and widely distributed in areas consisting of pyritiferous sediments or sulphide ores. Circulating subsurface waters containing sulphuric acid dissolve calcite. Gypsum is then precipitated in the form of free crystals or nodules embedded in shales, or forms druses of crystals in cavities in ore veins near the surface of deposits. In deserts free concretions of rose-shaped crystals called 'desert roses' may be found. They contain up to 50 per cent of sand and are formed by evaporation of subsurface water containing dissolved sulphates.

Large deposits of gypsum are found in the USA and Canada, and in Europe (in Great Britain, France, FRG, GDR, Poland and Czechoslovakia). The white, compact massive form called alabaster is found in Volterra, Italy, and in Spain. Burned gypsum, ground to a white powder, is known as Plaster of Paris and is used in many different ways.

Anhydrite is closely associated with gypsum, from which it forms by dehydration. It also occurs in hydrothermal deposits, e.g. in Andreasberg, FRG; and Cavnic, Romania.

3

Gypsum crystals are tabular (1, 5), columnar (6), or lenticular (2), and sometimes reach a length of more than a metre. Sometimes the mineral forms flat or spherical aggregates, and the crystals are often twinned in the form known as 'swallow-tail' (7). Aggregates are usually coarsely tabular, composed of parallel

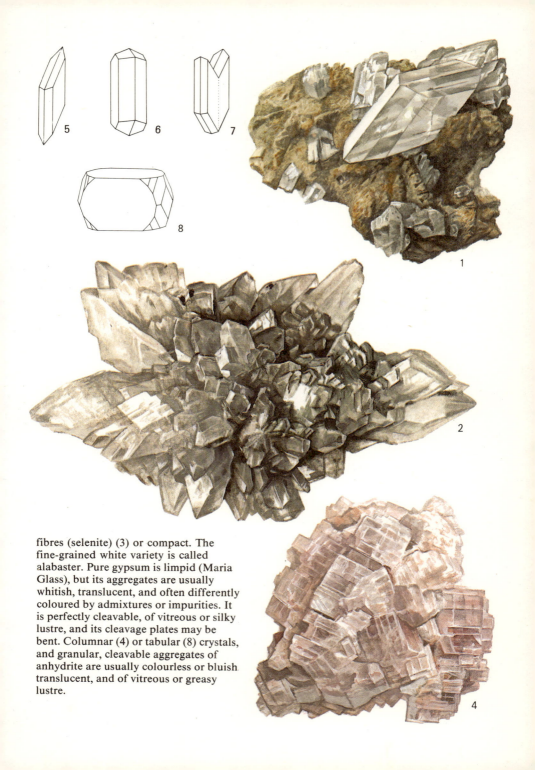

5   6   7

8

1

2

fibres (selenite) (3) or compact. The
fine-grained white variety is called
alabaster. Pure gypsum is limpid (Maria
Glass), but its aggregates are usually
whitish, translucent, and often differently
coloured by admixtures or impurities. It
is perfectly cleavable, of vitreous or silky
lustre, and its cleavage plates may be
bent. Columnar (4) or tabular (8) crystals,
and granular, cleavable aggregates of
anhydrite are usually colourless or bluish
translucent, and of vitreous or greasy
lustre.

4

Coelestine (celestite) $SrSO_4$
Orthorhombic; H. 3—3.5; Sp.gr. 3.9—4

The name coelestine stems from the Latin *coelestus*, meaning heavenly, and some coelestine crystals really do have the blue colour of the sky. The mineral's ability to colour the flame crimson-red has been used since ancient times in the production of the so-called Bengal fires (Bengal lights), but for many centuries it found no other practical application.

Coelestine is not very common but in some areas it may be quite abundant. It occurs in fissures and cracks in volcanic rocks, e.g. near Vicenza, Italy. It is found in smaller amounts in hydrothermal ore veins, e.g. in Leogang near Salzburg, Austria, and Špania Dolina, Czechoslovakia; the latter is a well-known occurrence of blue crystals. Usually, however, it originates as a result of complicated mineralization processes affecting sedimentary limestones, dolomites, and gypsum. As isomorphous admixtures these rocks and minerals contain trace amounts of strontium and barium. In the course of the decomposition of the rocks both elements are released and crystallize from circulating solutions in the form of coelestine and barytes, together with the gypsum, calcite, or native sulphur. The free crystals of coelestine are embedded in clay or form druses in cracks and vesicles in sedimentary rocks. In Europe they are found in the sulphur deposits in Sicily; near Tarnobrzeg, Poland; at Dornburg and Rüdersdorf, GDR; and at Clifton, Great Britain. Coelestine is recovered in Mokkatam, Egypt; and also from near Lake Erie, Michigan, USA, where the largest blue crystals have been found. Rich deposits of druse crystals were discovered recently in Libya.

2

148

1

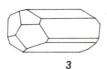

3

Crystallized coelestine has the form of plates and columns (1, 3). Usually it occurs in granular, stalky, or compact aggregates, sometimes of dish-shaped separation. Coelestine also forms plates composed of transversely parallel fibres or slender needles, reniform crusts, and irregular nodules. It is perfectly cleavable, aggregates have a conchoidal fracture, and the lustre is vitreous to pearly. Crystals are usually colourless and clear, or blue and transparent (2), rarely greenish or reddish. Aggregates are whitish, grey, or bluish, and are translucent. Similar minerals are barytes, aragonite, and strontianite. At present coelestine is used in the pyrotechnics, chemical and foodstuff industries, and in pharmaceutics. It is also a source of strontium, and exceptionally it is also cut as a gemstone.

149

Barytes (barite, heavy spar) BaSO₄

Orthorhombic; H. 3—3.5; Sp.gr. 4.48

Some minerals are named after one of their characteristic properties; one of these is barytes, which has its name derived from the Greek *barys*, meaning heavy. It was familiar even to early miners for its heavy weight, but for a long time they took it only for a useless gangue in ore deposits. It was classified as late as the beginning of the 17th century as an inorganic mineral which showed luminosity in the dark when heated. Thus, thanks to barytes, the luminescence of minerals was discovered.

Barytes originates in nature in many different mineralization processes, but it is nearly always formed by deposition from solutions. Only rarely has it crystallized at the time of cooling of eruptive rocks and pegmatites, e.g. in Alnö, Sweden. Hydrothermal barytes occurs in ore veins either as their only filling or associated with many other minerals. The latter type of deposit is most common. Such accumulations originated from hydrothermal solutions by the replacement of limestones, e.g. in rich deposits in Missouri, USA. Barytes, when it is found as a constituent of sedimentary rocks, is of a much more complicated origin and in part is formed in the course of processes described for coelestine. The sources of barium in this case were calcium carbonate and gypsum. Another source may be decomposed calcareous feldspars or orthoclase, in which barium forms an admixture. Barytes concretions, nodules, and impregnations are produced in this way in sediments, and may form large accumulations, e.g. the deposit in Meggen, FRG.

2

Barytes appears in more than 250 crystalline forms. Crystals are usually tabular (1, 3, 4), coffin-shaped, or columnar (5), and often form parallel twins (2). Aggregates are granular, fibrous, radiate, oolitic, reniform, or earthy. They form free concretions, nodules, rose-shaped aggregations, or stalactites, and may have dish-shaped separation. Barytes is perfectly cleavable, has glassy or pearly lustre, and is translucent or transparent. It is often colourless, clear, or milk-white, or may be of grey, yellow, or blue to reddish shades due to admixtures. Similar minerals are coelestine, aragonite, and fluorite. Barytes is widely used in ceramics (glazes, enamels), in the chemical industry (paints), as a filler in glossy paper, as an aid in well-drilling, in pyrotechnics, in medicine, and in special concretes for protection against radiation.

151

Crocoite PbCrO$_4$
Monoclinic; H. 2.5—3; Sp.gr. 5.9—6; up to 64 per cent Pb

Crocoite is a conspicuous heavy mineral of yellow-red colour, which was found for the first time in 1740 in the Urals and was called 'red lead ore'. In 1797, the French chemist Vauquelin discovered a new element, chromium, in it, and fixed its chemical formula. Crocoite got its name from the Greek *krokos*, crocus, because of its saffron-yellow colour.

Crocoite is neither very common nor a widely distributed product of weathering of primary lead ores, such as galena. It originates in the oxidation zone of their deposits only if the circulating subsurface water contains ions of chromium. Primary chromium ores (chromite), however, never crystallize with primary ores of lead. The source of chromium for crocoite is usually the dark basic intrusive rocks, and serpentines originating from them and containing disseminated chromite. Consequently, crocoite deposits are located in areas where polymetallic ores penetrate these intrusive rocks and serpentines. Associated minerals include pyromorphite, cerussite, wulfenite, vanadinite, vauquelinite, and limonite.

There are only a few localities of crocoite, e.g. Beresovsk, Nizhni Tagil, and Mursinsk, USSR; Dundas, Tasmania; Congonhas do Campo and Goyabeira, Brazil; and Penchalonga Mine and Umtali, Zimbabwe. Many-faced crystals are found in the gold mine near Labo, in the Philippines. Crocoite occurs rarely in Baiţa, Romania, and was discovered in 1977 near Callenberg, Saxony, GDR.

3

Crocoite usually has the form of many-faced (2, 3), slender columns (4) or acicular, longitudinally striated crystals. It forms large clusters (1) and druses (2) in cavities. It also occurs in the form of granular aggregates, pulverulent coatings, or inclusions. It is cleavable, breaks with uneven fracture, is brittle, and is of greasy or adamantine lustre. It is yellow-red with an orange streak, and is translucent. Similar minerals are realgar and cinnabar.

Crocoite is too rare to be of practical importance, but it is an attractive mineral and as such it is valued in every private or museum collection of minerals.

4

1

2

Scheelite CaWO$_4$

Tetragonal; H. 4.5—5; Sp.gr. 5.9—6.1; up to 80 per cent WO$_3$

Scheelite is an admixture of tin ores which has been exploited from the Bohemian and Saxon Ore Mountains since the Middle Ages. In 1781 the Swedish chemist K. W. Scheele discovered the element wolfram in it. Prior to this, the mineral was considered an iron ore, so-called tungsten (heavy stone), and miners called it by mistake 'white tin hailstone'.

Scheelite occurs in high-temperature hydrothermal deposits of tin, wolfram, and molybdenum ores, and is found in the Ore Mountains, Czechoslovakia and GDR; Bretagne, France; and Cornwall, Great Britain; as well as outside Europe. It also originated by the alteration of wolframite. Scheelite of metasomatic origin is found, for instance, in the deposit in Oruro, Bolivia. Exceptionally it formed in fissures associated with minerals of the so-called Alpine paragenesis. A large deposit of the stratiform type was recently discovered in crystalline schists near Mittersill, Austria.

Scheelite crystals usually occur isolated, in pointed-pyramidal form (1), and are seldom tabular (2). Aggregates form granular or reniform crusts. The mineral is cleavable, brittle, translucent, coloured grey-white, yellowish, brown or reddish, or, rarely, is colourless. The lustre is greasy or adamantine. In ultraviolet light it usually shows a light blue luminescence. It is an important tungsten ore.

2

1

154

Wulfenite PbMoO$_4$

Tetragonal; H. 3; Sp.gr. 6.7—6.9; up to 56 per cent Pb
and 39 per cent MoO$_3$

The weathered parts of deposits of primary lead ores contain certain
secondary minerals, such as wulfenite. It was first discovered in Bleiberg, Austria, and in 1785 was included in the mineralogical system.
Wulfenite is rare in nature, probably because the absence of molybdenum in solutions circulating through the oxidation zone of polymetallic ores is caused by a small participation of molybdenite in
this ore association.

The most important occurrences of wulfenite in Europe are Bleiberg, Austria; Mežica, Yugoslavia; Baiţa, Romania; and Příbram,
Czechoslovakia. Fine crystals are found in Tecomah, Utah, and Trigo
Mountains, Arizona, USA; Villa Ahumada, Mexico; Algeria; Moroc-
co; and Australia. Wulfenite in larger accumulations is a rich lead
and molybdenum ore.

2

Crystallized wulfenite has the form of
pyramidal columns (2), plates, and
cellular lamellae (1). Aggregates are
granular or compact. It is yellow,
honey-brown, or sometimes orange-red,
has a whitish streak, and greasy or
adamantine lustre. Wulfenite is perfectly
cleavable, breaks with uneven fracture, is
translucent or transparent, and brittle.

1

Apatite $Ca_5(PO_4)_3(F,Cl,OH)$
Hexagonal; H. 5; Sp.gr. 3.16—3.22

Apatite is a mineral of many different forms and various colour shades. It has long been mistaken for other minerals such as beryl, quartz, tourmaline, nepheline, and their varieties. At the end of the 18th century, however, A. G. Werner of the Bergakademie in Freiberg, Germany, discovered that it was, in fact, a new mineral and derived its name from the Greek *apatao*, meaning I am misleading.

Apatite is a very common mineral. Based on their origins, two different varieties may be distinguished. Crystalline and crystallized apatite is linked with magmatic activity or rock alteration, whereas apatite's amorphous and cryptocrystalline variety, called **phosphorite**, is of sedimentary origin. Apatite is a constituent of almost all igneous rocks, but it occurs in large quantities mainly in nepheline syenites in the Kola Peninsula, USSR. It is also contained in various pegmatites, e.g. in Arendal, Norway; and Renfrew, Canada. In the latter deposit crystals 3 m long and weighing several hundred kilogrammes were found. Apatite is also a constituent of hydrothermal deposits of tin ores, e.g. in the Ore Mountains (Czechoslovakia and GDR), and of iron ore deposits (e.g. Gellivare, Sweden). It is also represented among the minerals of the so-called Alpine paragenesis. Phosphorite forms small nodules, fibrous lenticles, and earthy intercalations in various sediments. Egg-shaped concretions tens of centimetres long come from the area around Podolsk, USSR.

4

Apatite crystals are many-faced, short- (1, 3) or long- (4) columnar to acicular, and attached or embedded. Aggregates are granular, radiate or compact, and have a dish-shaped separation. The mineral has many different colours, especially greenish, blue, violet, yellow, rarely pink, or reddish; sometimes it is even colourless and clear. Its lustre is vitreous to greasy, it is transparent, translucent, or opaque, and it has an uneven fracture. Some apatites fluoresce in ultraviolet light. Apatite is used in metallurgy and in the manufacture of fertilizers.

Phosphorite in nodules and egg-shaped
concretions of grey-black colour (2) is an
important raw material for the
production of phosphatic fertilizers.
Stalactitic crusts are called staffelite, and
many other varieties of phosphorite are
known under many different names.

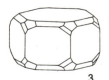

Pyromorphite $Pb_5(PO_4)_3Cl$

Hexagonal; H. 3.5—4; Sp.gr. 6.7—7; up to 75 per cent Pb

Minerals originating by the weathering of lead sulphide ores are usually varicoloured. Pyromorphite is an abundant constituent of secondary minerals found in the oxidation zone of polymetallic deposits. Its name was derived from the Greek *pyros,* fire, and *morphe,* form, because it was wrongly believed to be a product of volcanic activity. It appears in association with cerussite, and less frequently with mimetesite, in combination with other products of decomposition of primary ores of non-ferrous metals.

In countries with a long tradition of mining, pyromorphite occurrences are rare at present, because its subsurface accumulations have been depleted. On the other hand, newly discovered polymetallic deposits of pyromorphite and some other secondary minerals have become a rich source of lead. The variety with a higher content of calcium is called miesite (after the Stříbro-Mies deposit, Czechoslovakia). On the weathered ore it forms radiating, needle-shaped, almost spherical aggregates.

Typical deposits in Europe are found in areas containing polymetallic ores, e.g. Freiberg, GDR; Clausthal, Dornbach, FRG; Příbram, Stříbro, and Harrachov, Czechoslovakia; Poullaouen, France; Leadhills (Scotland) and Cornwall, Great Britain; and Nertschinsk and Beresovsk, USSR. It is also found in Phoenixville, Arizona, USA; Kabwe, Zambia; and in Australia and elsewhere.

1

Crystallized pyromorphite is very common. It occurs in the form of columnar (3) or needle-shaped crystals; plates are rare. Aggregates form granular, reniform, mamilliform, or botryoidal crusts and pulverulent coatings. Pseudomorphs after galena or cerussite are common. Most often it is green ('green lead ore' — 1), or brown ('brown lead ore' — 2), orange, yellow, white, or colourless. It is brittle, non-cleavable, of conchoidal fracture, of greasy to adamantine lustre, and has a grey-white streak. The less common mimetesite looks very similar.

Lazulite $2AlPO_4 . (Fe,Mg)(OH)_2$
Monoclinic; H. 5—6; Sp.gr. 3.1

Turquoise (callaite) $CuAl_6(PO_4)_4(OH)_8 . 5H_2O$
Triclinic; H. 5—6; Sp.gr. 2.6—2.8

Lazulite and turquoise have both been popular precious stones for thousands of years. Turquoise from deposits in the Sinai Peninsula was first recovered by the Ancient Egyptians, and then, in the Middle Ages, by the Turks. Lazulite was considered only an ornamental stone, and by the Arabs, in particular, it was used as an imitation of the precious turquoise and of the silicate lazurite called *lapis lazuli.* Its name is derived from the Arabic *azuli,* sky. Lazulite is found in pegmatites, in quartz veins, and in fissures in crystalline schists rich in aluminium; sometimes it is also found as a constituent of quartzites. Most important occurrences in Europe are in the Alps, e.g. in Werfen and Krieglach, Austria; Zermatt, Switzerland; and it is also found around Wermland, Sweden; and Zobor, near Nitra, Czechoslovakia. It also occurs in the elastic mica schists (itacolumites) in Crowders Mountains, North Carolina; the most perfect crystals come from Graves Mountains, Georgia, USA. Crystals from Tijuco, Minas Gerais, Brazil, are cut as precious stones.

Turquoise forms veinlets in eruptive rocks and irregular intercalations in sandstones. The most important deposits are in Mt Ali Missai, near Maaden, and in Meshed and Nishapur, Iran; Karatube Mountains, USSR; Tourquois Mountains, Arizona, USA; South America; and Los Cerillos and Mt Chalchuitl, New Mexico (where it was recovered by the Aztecs). In Europe it is found in Jordanów, Poland, and Plauen, GDR.

3

Lazulite in the form of perfect, pointed-pyramidal (1, 3), or tabular crystals is rare; it most often occurs in compact aggregates of uneven fracture. It is of deep blue, or sky-blue to blue-white colour, whitish streak, and vitreous lustre. It is used in making jewellery.

Turquoise usually forms compact aggregates (2), veinlets, granules, reniform crusts, and stalactites. Its colour is sky-blue, passing sometimes gradually into green-blue or grey-greenish. When exposed to the light or when affected by acids turquoise loses its colour and turns pale. It has a waxy lustre and splintery fracture. It is often replaced by artificial turquoise in jewellery.

Vivianite $Fe_3(PO_4)_2 . 8H_2O$

Monoclinic; H. 2; Sp.gr. 2.6—2.7

Vivianite is not very common. It originates in the reducing milieu near the surface of the Earth, being produced by the decomposing activity of water solutions charged with phosphoric acid acting on primary iron ores (e.g. pyrite, pyrrhotite, siderite, or rock-forming minerals rich in iron). Phosphorus ions in magmatic or metamorphic rocks come from disseminated apatite or from pegmatite phosphates, e.g. triplite and triphylite. The phosphorus contained in vivianite which has crystallized in vesicles of various fossils or impregnated fossil bones, or which occurs in peat, ferruginous clays, slates, and in the peat iron ore, is of organic origin.

Occurrences containing small amounts of vivianite are common, larger accumulations are rare. Perfect crystals come from St Agnes, Cornwall, Great Britain. In Tavistock, Great Britain, and Bodenmais, FRG, it originated on pyrite and pyrrhotite, whereas in Chvaletice, Czechoslovakia, it appears in manganese-pyrite slates. Vivianite crystallized in fissures is found in Kercz, Crimea, USSR; and Mullica Hill, New Jersey, USA. Crystals of extremely large size are found in Llallagua and Morococala, Bolivia, and in the recently discovered occurrences in Anlóua, Cameroon. Vivianite has no practical application.

Vivianite forms clusters of columnar and tabular crystals (3) or free, long-lenticular individuals which in Cameroon reach the length of more than a metre. Usually it forms radiate, fibrous, or stellate aggregations clustered in flat druses and reniform up to globular formations. It also occurs abundantly in earthy and pulverulent forms (2). Cleavage is perfect,

2

3

1

thin lamellae are flexible, and it is brittle and crumbles easily.

On fresh fracture vivianite is light to indigo-blue and transparent but it quickly becomes clouded and attains a greenish tinge (1). This is due to the oxidation of bivalent iron to the trivalent variety. The lustre is vitreous to pearly, and the streak is blue-white.

Wavellite Al$_3$(OH)$_3$(PO$_4$)$_2$. 5H$_2$O
Orthorhombic; H. 3.5—4; Sp.gr. 2.3—2.4

Wavellite is not very common. Nevertheless it is a very decorative mineral occurring in some eruptive rocks, tuffs, and sediments, where it coats cracks and bedding planes. It is deposited from solutions containing dissolved phosphorus, circulating, for instance, through rocks with microscopically disseminated apatite. The original source of phosphorus in sediments may be dead organisms which release the element during the process of fossilization. The necessary aluminium comes from argillaceous (clayey) minerals. Associated minerals are cacoxene, variscite, and some other phosphates.

Wavellite was found for the first time at the end of the 18th century in decomposed granite near St Austell, Cornwall, Great Britain. It was named in honour of the English natural scientist and physician, W. Wavell. Shortly afterwards it was discovered in greywackes in rich occurrences in Czechoslovakia (e.g. Milina, Cerhovice, and Třenice). It is also found in Langenstriegis, GDR, and Dünsberg, FRG. Large deposits are found in Hellerstone, Pennsylvania, USA; Bolivia; and Brazil.

Wavellite has no great practical application. At present it is used as a raw material for paints; in the USA it was used in the past as a source of phosphorus in the production of matches.

3

Wavellite crystallizes in the form of longitudinally striated columns (3), and needles clustered in radiate-fibrous, reniform, or globular aggregates. In fissures it forms flat radiate, stellate crusts (2) with fibres up to several centimetres long. Wavellite is most often colourless, whitish, yellowish, or green (1), but may even be brown or blue, and is sometimes concentrically banded. The streak is white and the lustre vitreous to silky. The mineral is translucent and perfectly cleavable. Aggregates affected by the decomposition into kaolinite (kaolinization) caused by weathering are usually white and earthy.

Destinezite (diadochite) $Fe_2(PO_4)(SO_4) \cdot 5H_2O$
Triclinic; H. 3; Sp.gr. 1.9—2.3

Delvauxite $2Fe_2PO_4(OH)_3 \cdot nH_2O$
Amorphous; H. 2.5; Sp.gr. 1.8—3.0

One of the products of pyrite decomposition, destinezite originates in some deposits of sedimentary iron ores or in schists containing pyrite and marcasite, but only if there is sufficient concentration of phosphate ions in the circulating solutions. The phosphorus comes from the bodies and shells of dead organisms. Destinezite associated with delvauxite, gypsum, vivianite, alunogen (keramohalite), aluminite, natrojarosite, and other decomposition minerals, originated in the limonitized iron cap of pyrite accumulations.

Destinezite and the associated minerals occur in sedimentary rocks in many countries. A much visited deposit is 'Garndorf pits' near Saalfeld, GDR. Here, in abandoned alum schist mines, visitors can admire the 'caves of nymphs' with white and brown-red dripstones, curtains, and terrace-shaped cascades on the walls. Loose nodules, often of considerable size, are found in Huelgoat and Peychagnard, Bretagne, France; Visé and Vedrin, Belgium; and in pyrite slates in the environment of Prague, and in Chvaletice, Czechoslovakia.

Delvauxite originates in the same way as destinezite and occurs, apart from the above-mentioned deposits, in Leoben, Styria, Austria; and Berneau, Belgium.

Large accumulations of both minerals are exploited as phosphate raw material. They have no other practical application.

Destinezite forms reniform or botryoidal aggregates, and more often earthy nodules (1) which are very similar to cauliflowers in shape. They are composed of six-sided plates of microscopic size and are usually the result of recrystallization of a colloidal amorphous substance, the so-called diadochite.

Destinezite nodules are usually friable, of varying size, and sometimes reach 50 cm in diameter. Destinezite is yellowish, brownish, rarely chalk-white, and of conchoidal fracture. Glassy varieties are usually translucent and of resinous lustre; earthy varieties are opaque.

Delvauxite nodules (2) are darker, and on a fresh fracture have a waxy lustre and a red-brown to chestnut colour.

2

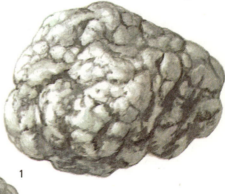

1

Torbernite (chalcolite) $Cu(UO_2)(PO_4)_2 . 8—12H_2O$
Tetragonal; H. 2—2.5; Sp.gr. 3.4—3.6; up to 61 per cent UO_3

Autunite $Ca(UO_2)_2(PO_4)_2 . 8—12H_2O$
Tetragonal; H. 2—2.5; Sp.gr. 3—3.3; up to 62 per cent UO_3

Torbernite and autunite, the so-called uranium 'micas', belong to the group of secondary minerals originating by weathering of primary uranium ores such as uraninite. Both these minerals, however, not only occur in uranium ore deposits but are quite common elsewhere. They are stable in the oxidation conditions of the Earth's surface. The weathering of magmatic eruptive rocks containing trace amounts of uranium, release the element which migrates in solutions and, under a sufficient concentration of phosphate ions and in the presence of dissolved copper, gets separated as torbernite in fissures of granite massifs, in greisens, pegmatites, and other rocks. In the presence of calcium it separates out as autunite.

Both uranium 'micas' were common in Jáchymov, Czechoslovakia. Miners, who have exploited silver ores at this location since the 16th century, called them 'green mica'. In 1797 the German chemist Klaproth proved that the newly discovered element uranium is also present in these 'micas'. Torbernite was named in honour of the Swedish chemist Torbern Bergman; autunite was named after the uranium ores deposit in Autun, France.

Uranium 'micas,' together with other secondary minerals, are quite common at the outcrops of all uranium deposits in Europe and elsewhere. In large accumulations they are exploited as uranium ores.

Torbernite crystallizes solely in the form of perfectly cleavable plates (1,3) and thin lamellae, often clustered in spathic aggregates. Isolated crystals may be several centimetres in size. The mineral is of deep green colour and light-green streak, and is translucent or transparent and of vitreous or pearly lustre.

Autunite is similar in form (4) and the two minerals are sometimes zonally intergrown. Autunite is yellow to yellow-green (2), of light yellow streak, and in contrast to torbernite it fluoresces strongly under ultraviolet light. Uranium 'micas' are significant prospecting indicators; coloured coatings of uranium micas on rock outcrops airtraced in a reconnaissance survey led to the discovery of rich deposits of uranium ores in the vicinity of Great Bear Lake, Canada.

1

3

4

2

Olivine (peridot, chrysolite) $(Mg,Fe)_2SiO_4$
Orthorhombic; H. 6.5—7; Sp.gr. 3.3

Olivine is an important constituent of some basic igneous rocks. It forms almost monomineral bodies of peridotites and dunites, and with pyroxenes and other minerals makes up part of dark gabbros, diabases, melaphyres, and basalts. Its grains form in metamorphosed dolomites and limestones under high temperatures and great pressure. It is also plentiful in some stony-iron meteorites.

Olivine is a mineral compound of two silicates — fayalite (Fe_2SiO_4) and forsterite (Mg_2SiO_4). Since it forms from magma at high temperatures, it contains trace amounts of nickel, replacing magnesium ions in the crystal lattice. The admixture of nickel is positively reflected in the alteration of olivine into serpentine in the process called serpentinization: magnesium separates in the form of compact magnesite, iron oxidizes into haematite and limonite, silicon turns into opal, and the released nickel forms aqueous nickel silicates, e.g. garnierite and pimelite. In this way large deposits of nickel ores have been formed in areas rich in serpentines (e.g. in New Caledonia).

Olivine occurs in many areas where geological conditions favour its origin. A well-known deposit of olivine of gem quality, called chrysolite (in Greek *topazios*) was worked in ancient times on the Zebirget Island, Red Sea. For a long time afterwards this site was abandoned, but it was re-discovered at the end of the 19th century. The gem chrysolite has been acquired in small quantities from the occurrence in Kozákov-Smrčí, Czechoslovakia, since the 16th century. At one time this was the only known deposit of chrysolite in Europe. Now, more important deposits have been found in Norway.

2

1

3

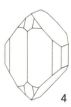

4

Olivine crystals have the form of prisms (3) or thick plates (4), which may be embedded or free, but usually it forms granules and large nodules enclosed in volcanic rocks (1). Cleavage is distinct, and aggregates have conchoidal fracture. It is translucent and glassy, and on fracture has a greasy lustre. It is olive-green, yellow-green or yellowish, and is often coloured brownish or brown-red by iron oxidation. Colourless olivines are very rare.

Chrysolite as a gem variety of olivine is transparent and pale green-yellow (2) in colour. Its lustre increases by cutting and polishing. Recently it has again become a popular ornamental stone.

171

Garnet group
Cubic; H. 6—7.5; Sp.gr. 3.4—4.6

The garnet group embraces about six main mineral species and their varieties. Garnets are aluminosilicates and silicates of calcium, magnesium, iron, manganese, and sometimes chromium and titanium, differing only in the amounts of these elements and their grading over into one another. They originate under high temperatures and great pressure, and are quite common. They have a relatively high density as small volumes form in the system by crystallization under great pressure. Consequently, they also form from other minerals—which under the more extreme conditions, especially in the regional metamorphosis of rocks, are no longer balanced as regards their development stage. The parent rocks of garnets are basic igneous rocks, eclogites, gneisses, and skarn ores. However, individual garnet varieties have specific origins of their own.

The calcareous **hessonite** originates at the contact of limestones with granitic magma. Important deposits are in Žulová, Czechoslovakia; Ala, Italy; and Oravița, Romania.

The ferriferous **almandine** (Oriental garnet) usually occurs as a constituent in gneiss and mica schists. Fine, large crystals are found in Zillertal, Austria, and Fort Wrangel, Alaska. Almandines of gem quality are found in Ceylon and India.

Pyrope contains magnesium and is formed in peridotites and kimberlites, and after serpentinization becomes a constituent of serpentines. Gem-quality pyrope is found in Třebenice and Podsedice, Bohemia, and therefore is called 'Bohemian garnet'. This mineral also occurs associated with diamond in Kimberley, South Africa ('Cape rubies'), and in Mirnyi, USSR.

Garnet crystals, embedded or attached, form characteristic rhombododecahedra (4), often many-faced (5, 6), sometimes weighing more than 10 kg. Aggregates are granular or compact. The fracture is uneven, splintery, and the lustre glassy to resinous; garnets are also dull, translucent or transparent, and of various colours. Hessonite (1) is hyacinth-red, grossularite greenish, almandine (2) violet-red, and pyrope (3,

7

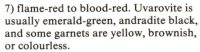

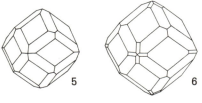

7) flame-red to blood-red. Uvarovite is usually emerald-green, andradite black, and some garnets are yellow, brownish, or colourless.

Free garnets (7) are usually found among the heavy minerals of river and sea sediments all over the world. Gem varieties of garnet are highly prized for use in jewellery.

Andalusite Al_2SiO_5
Orthorhombic; H. 7.5; Sp.gr. 3.1—3.2

Sillimanite Al_2SiO_5
Orthorhombic; H. 6—7; Sp.gr. 3.2

Disthene (kyanite, cyanite) Al_2SiO_5
Triclinic; H. 4.5 and 7; Sp.gr. 3.6.—3.7

This common aluminosilicate crystallizes in three different modifications formed in rocks rich in aluminium.

Andalusite occurs in pegmatites and crystalline schists. It forms large crystals in pegmatites, but more often it is found in masses composed of parallel slender laths of a length of several tens of centimetres. In crystalline schists it is usually disseminated, and especially in contact zones of granite massifs it achieves the form of long, thin, parallel fibres of cruciform shape, called chiastolite. When heated to high temperatures it grades into sillimanite and disthene. It is a widely distributed mineral. It was named after Andalusia in Spain, where it was found for the first time near Améria. Well-known occurrences are in the Alps, in the Urals, and in Minas Gerais, Brazil. It is mined in California, USA, and the USSR as raw material for the production of special porcelain and fire-proof materials.

Sillimanite is an accessory constituent in many rocks, most often forming wisp-like aggregates and felt-like asbestiform coatings.

Disthene occurs associated with andalusite and staurolite in crystalline schists, granulites, and eclogites. Fine specimens are found in Monte Campione, Switzerland; Passo di Vizze, Italy; and Bečov, Czechoslovakia; and workable deposits are in Wermland, Sweden; California, USA; Urals, USSR; and India. It is used for the same purposes as andalusite.

Crystallized andalusite has the form of columns (1) and lath-shaped crystals; aggregates are radiate-fibrous. Cleavage is imperfect, fracture uneven. It has glassy lustre and displays many different colours, most often pink, flesh-red, greyish, and pale violet. Sillimanite, which forms grey-white, brownish, or green-grey finely fibrous, cleavable aggregates (3), is most stable of the three modifications. Tabular crystals of disthene (2,4 — the black crystals are staurolite) are usually transversely striated; aggregates are radiating and lamellar. Disthene is perfectly cleavable, brittle, translucent, and of pearly lustre. It is blue, whitish, yellow, or green-grey, and sometimes also colourless. The hardness ranges from 4.5 in the longitudinal direction to 7 in the transverse direction. This special property of disthene is indicated by its name derived from the Greek *dis*, two, and *stenos*, power.

1

3

2

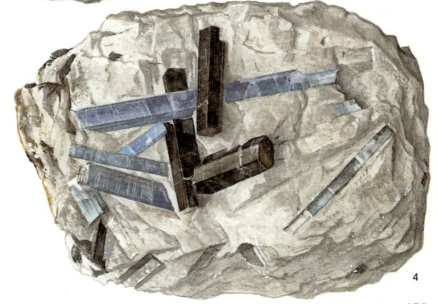

4

Topaz $Al_2SiO_4 (F,OH)_2$
Orthorhombic; H. 7.5—8; Sp.gr. 3.5—3.8

Topaz originates in intrusions of acidic intrusive rocks. It is found in some pegmatites, forms a substantial constituent of greisens, and has been separated out in the characteristic mineral association of tin-wolfram-molybdenum ores in quartz veins in and close to granite bodies. It also occurs disseminated as an accessory mineral in some kinds of quartz porphyry and trachyte. Under the effect of gas-charged hot solutions, some older rock-forming minerals may completely decompose. Topaz replaces them and forms large accumulations of the so-called topazites. This was the way in which the large topaz rock near Schneckenstein, GDR, was formed, and perfect pseudomorphs of topaz after orthoclase have been found here. The quartz porphyry in Mt Bischoff, Tasmania, was transformed in a similar way. Pycnite is the straw-yellow variety of parallel to radial, long lath-like aggregates of topaz found in ore veins associated with cassiterite and other minerals.

Topaz is a very common mineral, and mention can only be made of some of the most important occurrences. These include Minas Gerais, Brazil (crystals up to 300 kg); Volhynia, USSR; and Japan. Bluish crystals are found in Mursinsk, Urals, USSR; honey-brown crystals come from Villa Rica, Brazil. Topaz pebbles are found in wave-built placers in Ceylon.

4

3

176

Topaz crystals (1) occur in about 150 different forms. They are usually columnar (2, 4), many-faced, and longitudinally striated. Aggregates are acicular, granular, or compact. The mineral has a perfect transversal cleavage, conchoidal to uneven fracture, and is brittle, of glassy lustre, and translucent or transparent. Cleavage planes often display interference colours. Topaz is often colourless and clear but may also be coloured by admixtures to yellow, brownish, or reddish (2); it also occurs in bluish (3), pale violet, pink, and greenish colours. On exposure to sunlight its rich colours fade. Topaz is one of the oldest precious stones, and has remained popular and much prized in jewellery from the earliest times till today.

2

1

Staurolite $2Al_2SiO_5 . Fe(OH)_2$
Orthorhombic; H. 7—7.5; Sp.gr. 3.7—3.8

Staurolite had attracted much attention long before it was investigated as a mineral. It often occurs in the form of intergrown crystals or cruciform twins, which gave it its name from the Greek *stayros*, cross, and *lithos*, stone. Because of its unusual and conspicuous form it was worn as an amulet.

Staurolite is a typical mineral in some gneisses and mica schists; it also occurs in slates at the contact with intrusive rocks. It forms in rocks rich in aluminium. The associated mineral is usually disthene with which it often forms compound crystals.

Staurolite is found especially in areas built of crystalline schists. The fine, well-known specimens with blue disthene grown in the paragonite mica schist, come from Monte Campione, Switzerland, and from the Alps, e.g. Ritom and Vipiteno, Italy. Cruciform twins are found near Quimper, France, and there are several occurrences in the Jeseníky Mountains, Czechoslovakia, and near Aschaffenburg, FRG. Placer staurolites are found in the alluvium of the Sanarka River, Urals, USSR, and crystals of extremely large size are found in Gorob Mine, Namibia. New deposits of large but single crystals were found recently in New England, USA.

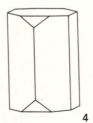

4

Staurolite occurs only in crystallized form; its short or elongated (4), perfectly terminated columns are embedded in the parent rock (2). More frequent than isolated single crystals are the characteristic diagonal interpenetration twins of two or three columns. The form known as the iron cross (1, 3) is typical of staurolite.

Staurolite has a distinct cleavage, conchoidal or splintery fracture, glassy, greasy, or dull (on the surface) lustre, and is opaque or translucent. It is brown, red-brown, or brown-red in colour, and the streak is white. Staurolite is of no practical importance, but it provides a good example of specific crystallographic peculiarities in the mineral kingdom.

1

3

2

Titanite (sphene) $CaTiSiO_5$
Monoclinic; H. 5—5.5; Sp.gr. 3.4—3.6; up to 41 per cent TiO_2

Titanite originates in many different ways. It is separated out during magmatic processes and occurs as one of the products of rock metamorphism. It crystallizes from solutions in cracks in crystalline schists as one of the minerals of the 'Alpine paragenesis'. Consequently, it is widely distributed. Its origin may be indicated by its colour; brown varieties most often occur dispersed in intrusive rocks and volcanites, and the green variety is more commonly found in the fillings of cracks. The variety called leucoxene, formed by the alteration of ilmenite or rutile, coats their grains with a grey-white film.

Some intrusive rocks contain large quantities of disseminated titanite, which under favourable circumstances forms extensive accumulations, e.g. massifs of alkaline syenites in the Kola Peninsula, USSR, where titanite originated in association with apatite, nepheline, and other rare minerals. The rocks there contain up to 16 per cent titanite, which is acquired as a by-product in working apatite. Titanite occurs in syenites near Dresden, GDR, and in volcanites near Laacher See, FRG. It is present in crystalline limestones near Boltano, Massachussetts, USA; in magnetite deposits near Arendale, Norway; and at Till Forster, New York, USA, and Renfrew, Canada. The most widespread resources of crystallized titanite, however, are in occurrences of minerals of the 'Alpine paragenesis', e.g. in Zillertal, Austria; Tavetsch, Switzerland; Passo di Vizze, Italy; Sobotín, Czechoslovakia; and Achmatovsk, USSR.

4

2

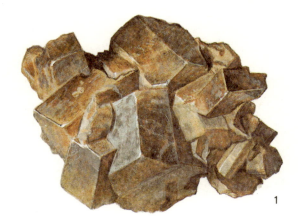

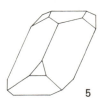

5

1

Titanite crystals are very common. Embedded crystals have the form of wedge-shaped envelopes (4); attached crystals are tabular (5) or columnar. Granular aggregates are rare. Titanite has an imperfect cleavage, is brittle, of conchoidal to dish-shaped fracture, and is transparent, translucent, or opaque, and of resinous to adamantine lustre. It is brown (1), yellowish, reddish, or grey-black, and the variety called sphene is green (2, 3) to brown-green. Titanites of good quality are much in demand for jewellery.

Larger accumulations of titanite are worked as an important ore of titanium. This light metal is used in modern industry in the manufacture of paints, abrasives and fireproof light alloys.

3

Hemimorphite $Zn_4(Si_2O_7)(OH)_2 . H_2O$
Orthorhombic; G. 4.5—5; Sp.gr. 3.3—3.5; up to 54 per cent Zn

The name hemimorphite is of Greek origin and is based on the mineral's semiform crystals, i.e. the self-same crystal is terminated at opposite ends by two different forms. Hemimorphite was identified as a new mineral and placed into the mineralogical system as late as 1853. Old miners, working the polymetallic deposits, knew it long ago as part of a mixture called galmei, composed of secondary zinc minerals, such as smithsonite ('carbonate calamine'), hydrozincite, hemimorphite ('siliceous calamine'), and willemite. It usually originates in the oxidation zone of polymetallic zinc and lead deposits.

Hemimorphite does not only crystallize in the so-called iron hat of ore deposits. Large accumulations mixed with other secondary lead and zinc minerals were formed by metasomatic replacement of limestones and dolomites. This mixture is a rich ore worked in large deposits in Bleiberg, Austria; Rabelj, Yugoslavia; Ołkusz, Bytom, and Tarnowice, Poland; and near Vieille Montagne and La Calamine, Belgium (hence the name 'calamine'). Important deposits of hemimorphite are found also in Sardinia; Matlock, Great Britain; and especially in Djebel Guergour, Algeria, and Leadville, Colorado, USA. Large and beautiful crystals also come from Lavrion, Greece, and Nertschinsk, Transbaikal, USSR.

3

Crystallized hemimorphite has the form of elongated plates (1) with one flat and one wedge-shaped termination (3), with which they usually grow to the base. They are often flabelliform. Aggregates are usually granular, finely fibrous, of agate-like texture (2), concentrically banded, and with reniform surface. They form encrustations, coatings, and stalactites. Hemimorphite is cleavable, brittle, of conchoidal fracture, and translucent or transparent. It is variously coloured according to admixtures; it may be whitish, yellowish, greenish, light blue, or sometimes colourless and clear, and of a full glassy up to adamantine lustre. The

1

larger accumulations of hemimorphite
are worked as an important zinc ore. The
polarity of its vertical axis is the cause of
a pyroelectric phenomenon; when it is
slightly heated, the tops of its crystals
become electrically charged.

2

Epidote (complex aluminosilicate with calcium and iron)
Monoclinic; H. 6—7; Sp.gr. 3.3—3.5

Epidote is an important example of a mineral formed by the metamorphism of rocks. It originates at relatively low temperatures and high pressures, especially in dolomitic limestones, altered to erlans and skarns, or at the contact with the magma. Associated minerals are garnet-hessonite, vesuvianite, amphibole, pyroxene (fassaite, hedenbergite), magnetite, and others. It was also formed in the regional metamorphism of dark igneous rocks such as gabbros. Its almost monomineral accumulations are called epidosites.

A very common kind of epidote is the one resulting from thermal decomposition and weathering of rock-building silicates rich in calcium, e.g. plagioclase-feldspars, amphiboles, pyroxenes, and scapolite. As a younger mineral, it occurs in cracks in various intrusive rocks, in gneisses, amphibolites, and chloritic schists. It is usually perfectly crystallized and forms rich druses on the walls of cavities. Its bunches, thickly set with hairs and fine fibrous aggregates, are called pistacite.

Epidote is widely distributed and common. The most beautiful crystals came from the deposit in Knappenwand, Sulzbachtal, Austria. This deposit, however, has been commercially unimportant for decades now. Perfect crystals are found in Achmatovsk, USSR, and Arendal, Norway. Other deposits are in Zermatt, Switzerland; Strzegom, Poland; Sulzer, Alaska; and in the copper-bearing deposits near Lake Superior, Michigan, USA.

2

3

184

1

Epidote crystals are usually many-faced (2), and some 200 different forms are known including columnar, lath-like (3), acicular, always longitudinally striated. They form bunches of parallel pectiniform aggregations (1), grown with their sides to the walls of flat cavities. Aggregates are stalky, radiate, granular, or massive. Epidote has a distinct cleavage, conchoidal up to splintery fracture, a full glassy lustre, and a grey streak. It is transparent, or, more often, translucent. Crystals are dark green to black-green; pistacite is light green. Occasionally a reddish form of epidote occurs, e.g. in Zillertal, Austria, and Glencoe, Scotland. Similar minerals are actinolite, greenish vesuvianite, pyroxene-fassaite, and tourmaline.

Vesuvianite
(basic aluminosilicate with calcium, magnesium, and iron)
Tetragonal; H. 6.5; Sp.gr. 3.27—3.45

Vesuvianite was named after its occurrence on Vesuvius, Italy. It is a typical metamorphic mineral originating primarily by the alteration of limestones and dolomites at the contact with igneous and volcanic rocks. Its grains are usually embedded, and crystals form rich druses in fissures and cracks in carbonates and crystalline schists. This specific association of contact minerals includes garnets (hessonite, grossularite), wollastonite, diopside, clinochlore, and other calcium silicates. Vesuvianite occurs in many differently coloured varieties, most of which have local names. It was originally erroneously confused with garnets, tourmaline, olivine and topaz, and it was not identified as a separate mineral until 1795 when it was determined by A.G.Werner.

A typical occurrence of vesuvianite is Hazlov near Cheb (Eger in German), Czechoslovakia, where a dark to yellowish-brown variety of parallel columns, called egeran, is found. Vesuvianite of the same origin is found at Žulová, Czechoslovakia; Göpfersgrün, FRG; Cziklova and Dognecea, Romania; Monzoni, Italy; Achmatovsk, USSR; and Eiker, Norway. Crystals tens of centimetres long come from Magnet Cove, Arkansas, USA, and Morales, Mexico. Vesuvianites from Brazil and Kenya are used in jewellery.

1

186

2

3

Vesuvianite crystallizes in the form of columns (1, 3), often many-faced, and longitudinally striated. Aggregates are usually flabelliform (2) or composed of long parallel stalks, or granular or massive. Vesuvianite is not cleavable and breaks with uneven, splintery fracture, and it has glassy to greasy lustre, is usually opaque, but may also be translucent to transparent. Its typical colours are brown, yellow-brown, and green; it may also be black-brown. Blue vesuvianite, called cyprine, is found very rarely, e.g. in Souland, Norway, and pink crystals are known from Jordanów Śląski, Poland. Vesuvianite of good quality is cut as a precious stone.

Similar minerals are garnet (which is, however, never striated), zircon, and cassiterite. By decomposition, vesuvianite alters into mica, pyroxene, scapolite and garnet.

187

Beryl and its varieties $Be_3Al_2(SiO_3)_6$

Hexagonal; H. 7.5—8; Sp.gr. 2.6—2.8; up to 14 per cent BeO

Emerald and aquamarine, the precious varieties of beryl, have been known since ancient times. Common beryl is the product of magmatic activity, and it usually occurs in granite pegmatites. Gas-charged hydrothermal solutions favoured the origin of large crystals. In the deposits in Acworth and Crafton, New Hampshire, USA, beryls of up to 1.5 tonnes were found; in Albany, USA, a column 10 m long and 1.8 m in diameter was found; and in newly discovered deposits in Mozambique and Madagascar, a column of 18×3 m weighing 379 tonnes was found. Beryls formed in mica schists and limestones are of hydrothermal origin.

Beryl is quite common. In Europe it is found in Limoges and Chanteloube, France; Zwiesel and Tischenreuth, FRG; Meclov and Písek, Czechoslovakia; Finbo, Sweden; and Cornwall, Great Britain. Large deposits are also found in the USA and in South Africa.

Aquamarine is the light blue-green form of beryl. Its characteristic colour is due to the transition of electrons between ferrous and ferric ions in the trace amounts of iron. Aquamarine originates in pegmatites from solutions either directly or by a chemical decomposition of older beryls.

Common beryl forms columnar crystals (1, 4) several metres long, often longitudinally striated, of transversal cleavage, and embedded or attached. It also forms coarsely stalky aggregates. It is brittle, of uneven fracture, opaque to translucent, of grey-yellow to yellow-green colour, and of dull or greasy lustre. It alters by decomposition, giving rise to phenakite and bertrandite, and weathers into muscovite and kaolin. Beryl is an important raw material of beryllium, which is used in nuclear physics as a neutron moderator, and in

3

1

metallurgy as an ingredient in hard
alloys. Aquamarine crystallizes in slender
columns (2), which are usually corroded.
It is pale blue to blue-green, having the
colour of sea water (hence its name). The
lustre is glassy, and it is translucent to
transparent (3).

4

2

Beryl and its varieties (continued)

The most important deposit of aquamarine is in Theofilo Otoni, Minas Gerais, Brazil. The largest crystal known to have been found there (in 1910) weighed approximately 109 kg; in 1961 a crystal was found which was 73 cm long, 15 cm in diameter, and weighed about 21 kg (equivalent to 70,000 carats of gem raw material of top quality). Aquamarines are also found in Mursinsk, Adunczilon, and Takovaya River, Urals, USSR; San Piero, Elba; and the Mourne Mountains, Ireland.

Emerald is a precious variety of beryl containing a trace amount of chromium (up to 0.3 per cent). These trivalent ions in the crystal field of the space lattice cause the typical rich green colour of emerald by selective absorption of transmitted light. Emerald is separated out from hydrothermal solutions within magmatic bodies, and also occurs as grains disseminated in granite, e.g. in Eidsvald, Norway.

Emerald was exploited near Djebel Zabara in Egypt some 4,000 years ago. Rich deposits were discovered by the Indians in the 15th century near Muzo and Schivory, Colombia, in veins of white calcite, and the deposits are still worked at present. Emeralds are also found in mica schists in the deposits in Takovaya River, USSR, and Habachtal, Austria. Crystals of up to 25 cm come from the Transvaal, South Africa, and new deposits were discovered recently at Stoney Point, North Carolina, USA.

Morganite (vorobyevite) contains an admixture of caesium, and its fine pink colour is due to the admixture of manganese. Important occurrences are found in Marharitra, Madagascar; Pala and Katrina, California, USA; and Theofilo Otoni, Minas Gerais, Brazil.

3

Emerald (1) has a deep rich green colour, and transparent emeralds of perfect colour are sometimes more valued than diamonds of the same size. However, crystals easily become clouded due to a network of tiny microflaws. The mineral's glassy lustre increases when it is cut and polished. Crystals of rose-red morganite (2) are usually thick-columnar up to tabular, many-faced (3), and transparent or translucent. Beryls of deep red colour are rare, occurring, for instance, in rhyolites in the Wah Wah Mountains, Utah, USA. The green-yellow to golden opalescent form of beryl, known as heliodor, comes mostly from

1

Namibia. Colourless clear beryl is called
goshenite. Minerals similar to beryl and
its varieties are apatite, quartz,
tourmaline and its varieties, and topaz.

2

Tourmaline and its varieties (very complex basic borosilicate of sodium, aluminium, iron, magnesium, calcium, etc.)
Rhombohedric; H. 7—7.5; Sp.gr. 3—3.3

The unusual physical properties of some minerals are often discovered only by chance, and this was the case with those of a stone from Ceylon which was called *turamali* (i.e. stone attracting ash) by the native inhabitants. The stone exhibited a strange property: heated crystals attracted ash from the fire on which they had been heated. This phenomenon, later called pyroelectricity, originates on account of a special internal structure of crystals, in which an axis has no symmetrical ends and is therefore polar. When the crystal is heated, positively and negatively charged particles are present at the opposite ends of the axis.

Tourmaline is a typical mineral of acidic igneous rocks. Sometimes it forms directly in granites, but most of it crystallizes in granite pegmatites and metamorphic rocks from gas-charged hydrothermal solutions containing fluorine and boron. Consequently, it is also a contact mineral, a constituent in greisens and quartz veins with cassiterite, and occurs in some siderite veins. Tourmaline disseminated in mica schists originates by the metamorphosis of clayey sediments, mostly containing trace amounts of boron. There are many occurrences of tourmaline, yielding several differently coloured varieties. Black schorl is a common variety.

2

1

3

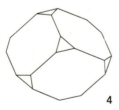

4

Tourmaline occurs in the form of longitudinally striated or channelled columns (1, 3), or needles with differently shaped ends on the same crystal. Flattened, lenticular crystals (2, 4) are rare. Crystals occur embedded as well as attached, sometimes reaching a length of up to 3 m, e.g. in Skrumetorp, Sweden. Aggregates composed of coarse stalks to fine fibres sometimes form flabelliform and radiate-fibrous 'tourmaline suns' in fissures. Tourmaline is brittle, non-cleavable, of conchoidal fracture, and has a glassy lustre. Varieties differ markedly in colour on account of admixtures. Schorl is pitch-black and opaque (1, 2). It is used in jewellery as a special polarizing apparatus called tourmaline tongs.

Tourmaline and its varieties (continued) ✓

The brown dravite with a relatively high content of magnesium occurs in mica schists near Dobrova, Yugoslavia, and was named after the Yugoslavian river Drava. Large, wide columns, limited on both ends by faces come from Yinniethara, Australia. Pink, green and blue varieties of tourmaline crystallize in lithium pegmatites and often occur combined in one deposit. Rubellite is the most frequent of these varieties, and in Europe it occurs in Rožná, Czechoslovakia; Penig, FRG; San Piero, Elba, Italy; and Mursinsk and Shaitansk, USSR. Brilliant crystals are found in the Pala and Mesa Grande occurrences, California, USA, and Madagascar; newly discovered localities occur in Mozambique. The most important occurrences of verdelite, the green variety, are found in Minas Gerais, Brazil, and the rare blue variety, indigolite, is found in Utö, Sweden; Goshen, Massachussetts, and Pala, California, USA; and in placers in Ceylon. The limpid colourless achroite forms large druses, especially in Elba, Italy. Here, also, the so-called 'Moor's or Turkish heads', formed of colourless to greenish columns with black tops, may be found.

The different colours of the varieties described above are due to the presence of manganese or iron in the crystal lattice. By selective absorption of transmitted light manganese causes the pink colour of rubellite, and iron the green colour of verdelite and the blue colour of indicolite.

2

3

194

1

Rubellite occurs crystallized as well as in the form of stalky aggregates (1), and it may be found in various shades of pink, and sometimes even red. It is usually translucent; however, transparent stones are highly valued in jewellery. Verdelite can be found in many shades of green, and the blue colour of indigolite ranges from sky-blue to deep azure-blue. Of a special brilliance are varicoloured crystals with concentric colour bands (2, 4) or transverse bands (3). Wonderful varicoloured precious varieties of tourmaline come from Pala Chief, San Diego County, California, USA, and from Minas Gerais, Brazil. They are mostly used in jewellery. Dravite (5) is usually brown, or green-brown to brown-black, and translucent. Minerals resembling tourmaline include some epidotes, amphibole, augite, apatite, beryl, and the varieties of quartz.

4

5

Dioptase $CuSiO_2(OH)_2$
Trigonal; H. 5; Sp.gr. 3.3; up to 50 per cent CuO

Dioptase is not a very common secondary product of weathering of primary copper ores. The conspicuous emerald-green colour of numerous crystals in cavities in ore veins attracted the attention of miners a long time ago. On account of its colour they called the mineral 'Kupfersmaragd' (copper emerald). It was included in the mineralogical system as late as 1797 by the French crystallographer R. J. Haüy under the name of dioptase.

Emerald and dioptase provide a good example of how two different causes can produce the same colour in two different minerals. The cause of the green colour of emerald is the admixture of chromium, in dioptase the cause is the admixture of copper. The effects of different crystal fields of the crystal lattice of these minerals on ions of chromium and copper resulted in an identical selective absorption of transmitted light, and they are both thus coloured emerald-green.

Dioptase occurrences are not very common. The mineral is found in calcite veins in Altyn Tybe, Kirghizia, USSR, and elongated crystals associated with wulfenite and hemimorphite come from Baiţa, Romania. Other occurrences are in Copiapoó, Chile; Peru; and Arizona, USA. The most important localities of beautiful and large crystals are in the vicinity of Otawi, Namibia, and in Zaire in copper deposits, where dioptase forms a constituent of the mined ore. In the Shaba province (formerly Katanga), columnar crystals suitable for cutting into gemstones are extracted.

2

Dioptase almost always occurs crystallized in large druses in cavities near the surface of copper deposits, or in cracks in calcite or dolomite (1). The crystals have the form of small, mostly short columns (2). Dioptase shows perfect rhombohedral cleavage, is brittle and of uneven fracture, and is usually translucent or transparent and of shiny, glassy lustre. It is coloured emerald-green to black-green, and the streak is green. Black dioptase is found in the Tiger deposit in Arizona, USA. Laboratory tests have shown that the unusual colour is due to the loss of water caused by a subsequent heating of crystallized dioptase to a temperature of some 700° C.

On account of its rare occurrence dioptase is of no practical use as a copper ore. Transparent crystals are used in jewellery, but their faces are relatively soft, and they soon lose their colour as well as their lustre.

1

Augite (very complex silicate of calcium, magnesium, aluminium and iron)
Monoclinic; H. 6; Sp.gr. 3.3—3.5

Augite belongs to the pyroxene family and is quite common and abundant in nature. It was originally mistaken for tourmaline. It was distinguished as a new mineral in the 18th century by A. G. Werner. It is found in rocks such as diabases, melaphyres, basalts and their tuffs. It also originates by contact metamorphosis. The most important occurrences in Europe are Černošín and Bořislav, Czechoslovakia; Kaiserstuhl, FRG; Auvergne, France; Vesuvius, Etna, and Stromboli, Italy; and Arendal, Norway.

Amphibole (complex silicate of calcium, magnesium, iron and titanium)
Monoclinic; H. 5—6; Sp.gr. 2.9—3.4

Amphibole is the name given to a whole group of rock-building silicates, and was derived from the Greek *amphibolos*, uncertain. Common amphibole is found in basic igneous rocks, as a contact mineral, or as the product of regional metamorphosis, when it formed a rock called amphibolite. A special type of amphibole occurs in basalts and their tuffs. The most typical free crystals come from Vlčák, near Černošín, and Kostomlaty, near Bílina, Czechoslovakia.

Actinolite (complex silicate of calcium, magnesium and iron)
Monoclinic; H. 5.5—6; Sp.gr. 2.9—3.1

Actinolite belongs to the amphibole family. It forms the rock constituent in crystalline schists or occurs as isolated embedded crystals. It also originates by contact metamorphism. It forms secondarily in igneous rocks by the alteration of augite or olivine, and being itself an unstable mineral it also changes into talc. It is very common, and characteristic crystals come from Zillertal, Austria.

Augite originates in magmatic rocks in the form of flattened columns terminated by wedge-shaped planes (1, 4). It is cleavable, brittle, opaque, of black colour, and green-grey streak. It often forms twins or granular nodules.

Amphibole crystallizes in the form of six-sided columns usually terminated by only three faces (2, 5). Aggregates are stalky or fibrous. It is cleavable, sometimes translucent, glassy, of black colour, (sometimes with a brown-green tinge), and the streak is dark grey.

Actinolite forms long-columnar crystals, and radiate to fibrous aggregates ('radiated stone' — 3). Compact aggregates composed of microfibres are called nephrite; felt-like asbestiform aggregates are known as byssolite. Actinolite is cleavable, glassy, opaque or translucent, and of green colour, though it is sometimes also grey-white.

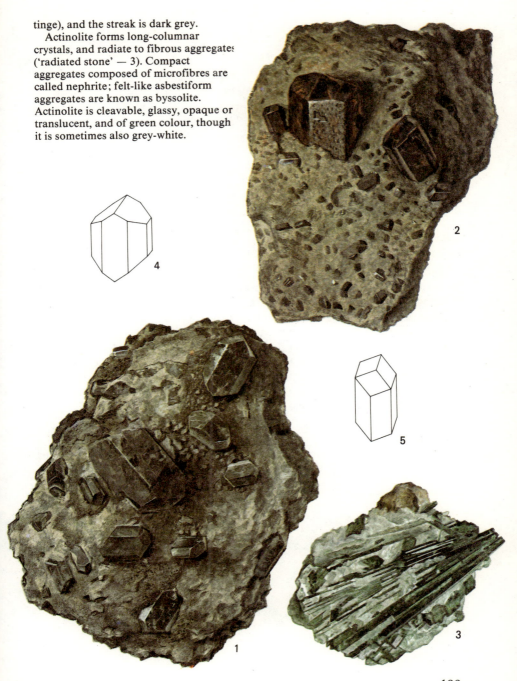

199

Spodumene LiAlSi$_2$O$_6$

Monoclinic; H. 6.5—7; Sp.gr. 3.1—3.2; up to 8 per cent Li$_2$O

Spodumene is a pyroxene which usually occurs in lithium pegmatites as a product of acidic granite magma. It is also found in limestones on their contact with granite. It forms large crystals. In the deposit Etta Mine, Black Hills, Dakota, USA, a flattened column measuring 13 × 1.8 × 0.9 m and weighing some 66 tonnes was found in about 1915.

Associated minerals are those occurring in pegmatites. Spodumene is not very common, its occurrence being limited to certain areas of granites. In Europe it is found in Vipiteno, Italy; Rauris, Austria; Utö, Sweden; Sobotín, Czechoslovakia; Scotland; and Ireland. Large deposits are found in the Black Hills, Branchville, Dakota, USA; Minas Gerais, Brazil; and Madagascar. It has two gem varieties which are described below.

Kunzite, coloured pink by the admixture of manganese, was discovered in 1903 in the Pala deposit, California, USA, and later also in Madagascar and Brazil. In Brazil, in 1961, the largest known crystal of kunzite, measuring 31 × 15 cm and weighing 7.5 kg, was found in the Urucupa Mine near Itambacuri, Minas Gerais. Large deposits of first-class kunzite are worked in Mawi and Lagleman, Afghanistan.

The green variety called **hiddenite** achieves its colour by the admixture of chromium. Its deposits include Stoney Point, North Carolina, USA; and Madagascar.

4

3

200

1

2

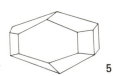

5

Spodumene often occurs in the form of long columns (4) with longitudinal striation, or it may be tabular (5) or spathic. Aggregates are stalky, fibrous, coarse spathic (1), or scaly. It is perfectly cleavable, has uneven fracture, and glassy to pearly lustre. It is yellowish, greenish, white or, rarely, colourless, and translucent.

Kunzite (2) is of a deep pink colour, sometimes with a violet shade. It shows phosphorescence in ultraviolet light. Hiddenite (3) is emerald-green to green-yellow. Both varieties turn grey when exposed to the air, yet they usually preserve their original colour in the core. Spodumene is an important raw material in the production of lithium.

Pectolite $Ca_2NaHSi_3O_9$
Triclinic; H. 5; Sp.gr. 2.7—2.8

Pectolite, found in occurrences typical of zeolites, was described in 1828 by the German mineralogist F. O. Kobell, from basalt bodies in Italy. It derives its name from the Greek *pectos,* folded, and *lithos,* stone. It is usually found in basic igneous rocks, e.g. melaphyres, or as a filling in cavities and veins, and is often associated with zeolites, which look very similar.

Its occurrences are in volcanic areas. Numerous fine specimens are found in Želechov, Czechoslovakia; Monte Baldo, Italy; Långbån, Sweden; and Ayrshire, Scotland. Other important locations are in Bergen Hill, New Jersey, USA; Canada; and Greenland.

Prehnite $Ca_2Al_2Si_3O_{10}(OH)_2$
Orthorhombic; H. 6—7; Sp.gr. 2.8—3

Prehnite is a secondary mineral formed by the decomposition of calciferous silicates. It occurs in fissures and cavities of basic igneous rocks (diabases and melaphyres) or in fissures in intrusive rocks (gabbros) and in crystalline schists, particularly amphibolites. It is very common, in some places being the most important constituent of decomposed rocks. Associated silicates, besides zeolites, include epidote and zoisite in deposits near Lake Superior, USA, and also native copper. It is found in Dillenburg and Harzburg, FRG; Val di Fassa and Vipiteno, Italy; Jordanów, Poland; and Markovice, Czechoslovakia. Large accumulations are found in Patterson, Bergen Hill, New Jersey, USA.

2

1

4

5

Crystallized pectolite (4) is rare; it occurs in the form of plates or columns. More typical are aggregates with fibres and needles radiately grouped in a star shape (1). Pectolite is cleavable, of glassy, pearly or silky lustre, and translucent to opaque. It is white to pinkish in colour, and sometimes colourless. Similar minerals are natrolite, tremolite, and wollastonite.

Prehnite crystals most often have the form of plates (5) and columns clustered in semi-globular forms (2). Fibrous aggregates form reniform crusts (3). Prehnite is cleavable, breaks with uneven fracture, is translucent, and has a glassy to waxy lustre. It may be colourless, but is usually grey-white with a greenish or yellow-green tinge. Similar minerals are wavellite, stilbite, heulandite, barytes, and aragonite.

3

203

Micas (complex aluminosilicates with iron, magnesium, etc.)

Micas are very common and widely distributed minerals forming a large family of many species and varieties. Some of them are important raw materials for different industries, and a source of important elements. Of the numerous mica varieties, muscovite, zinnwaldite, lepidolite, and biotite will be covered here.

Muscovite $KAl_3Si_3O_{10}(OH,F)_2$
Monoclinic; H. 2—2.5; Sp.gr. 2.78—2.88

Muscovite and biotite are the most common mica varieties. Muscovite was named after Moscow, USSR, from where large sheets of light mica found in the Urals were imported to western Europe. Muscovite occurs as the main constituent of many rocks, such as acidic igneous rocks and pegmatites, arises by the metamorphosis of rocks into mica schists and gneisses, and is a secondary mineral resulting from the alteration of many silicates. Large crystals are found in pegmatites in the Urals, USSR; USA; Canada; and especially in India. In the Inikurti Mine, Nollore, a monocrystal 4.5 metres long, 3 metres in diameter, and weighing 77.1 tonnes was found. Muscovite is chiefly used as an insulating material in electronics; in the past large paper-thin sheets were used as window panes, and in heat-proof windows of old stoves and ranges.

Zinnwaldite

This mineral is a mica variety containing lithium and iron. It crystallizes from gas-charged hydrothermal solutions with ores of tin, wolfram, molybdenum, etc. It is also a constituent in greisens. It was found for the first time in Zinnwald (Cínovec), on the frontier between Bohemia and Saxony. It also occurs in Altenberg, GDR, and Cornwall, Great Britain.

Lepidolite (lithionite)

This lithium mica containing no iron is a constituent of lithium pegmatites. It occurs associated with coloured tourmalines, petalite, amblygonite, and other minerals of this paragenesis. Lepidolite was discovered in Rožná, Czechoslovakia; it occurs also in the GDR; on Elba; in the Urals; in the USA; in Canada; and in Namibia. Large masses are mined in Antsirabe, Madagascar. It is a source of lithium and caesium.

Muscovite crystallizes in the form of six-sided plates (2, 3) and columns (1). Its scales occur isolated or form spathic, rose-shaped, scaly, finely granular, or massive aggregates (sericite). It cleaves perfectly, is flexible in thin plates, and is transparent to translucent, pearly, and colourless or yellowish. Zinnwaldite usually forms plates of parquet-like pattern arranged into flat druses (4). It is usually yellow-grey or greenish, with wrinkled planes, and is perfectly cleavable and glassy. Crystallized lepidolite is rare; it mostly forms thin-lamellar and finely scaled aggregates (5). It is pink, pale violet, greenish, or sometimes grey and whitish, and has a high lustre.

5

2

3

4

1

Biotite (complex basic aluminosilicate with potassium, magnesium and iron)
Monoclinic; H. 2.5—3; Sp.gr. 2.8—3.2

Biotite is a common mica found as a rock constituent in granites, pegmatites, dark eruptive rocks, and volcanic ejecta. It was also formed by metamorphosis in gneisses, mica schists, and contact hornfels. It easily alters into chlorite, and has many varieties. A similar mineral called phlogopite (a mica containing magnesium and fluorine) is a rock constituent in contact limestones and dolomites. As a primary mineral in pegmatites, it sometimes forms crystals of huge dimensions. An example is a columnar crystal with a diameter of more than 4 m and more than 10 m long (and weighing about 333 tonnes) which came from the deposit in Lacey Mine near Loughborough Township, Ontario, Canada.

Biotite is swidely distributed. Crystals have been found on Vesuvius, Italy; Pajsberg and Långbån, Sweden; Miass, Southern Urals, USSR; Templeton, Canada; and elsewhere.

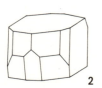

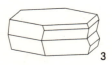

Pseudohexagonal columns of biotite (2) are embedded or attached (1), and often twinned (3). Aggregates are mostly scaly, spathic, and lamellar. It forms perfectly cleavable, flexible, and elastic scales and plates of vitreous to submetallic lustre, and it is translucent or opaque. It is dark brown to black; weathered specimens are bronze-coloured. It easily crumbles into soft earth. It has no practical application.

206

Talc $Mg_3Si_4O_{10}(OH)_2$
Monoclinic; H. 1; Sp.gr. 2.7—2.8

Talc is one of the minerals which has been known and used by man for thousands of years. The soft, fire-resisting raw material was formerly used in the production of pots or was carved into ornaments. It originates by alteration of magnesium minerals and rocks at low temperatures and high pressures. It occurs in chloritic schists, crystalline limestones and dolomites, and magnesites, and is found on their contact with igneous rocks, and in serpentines. The largest deposits are in the USA, Canada, Austria and Italy. It is used in the manufacture of fire-resisting ceramic articles for the electrical industry, and in the paper, cosmetic, chemical and textile industries.

Talc forms only aggregates of macroscopic scales or lamellae, which are often finely wrinkled (1). The compact variety is called steatite (from the Greek *stear,* suet). Larger scales and plates are flexible, but inelastic. It cleaves perfectly, is translucent or opaque, and is of pearly lustre, light greenish, yellowish, or grey-white colour, and greasy or soapy feel.

1

Chrysotile $Mg_5Fe(OH)_8Si_4O_{10}$
Monoclinic; H. 3; Sp.gr. 2.6

Fine fibrous chrysotile and scaly antigorite belong to the serpentine family. Chrysotile is an important mineral originating in serpentization, which is an alteration process of olivine rocks (peridotites), gabbros, and diabases. In the course of the process the original rock increases its volume, the body of the serpentine breaks irregularly, and the cracks and fissures are filled with parallel-fibrous chrysotile (asbestos) with fibres at right angles to the edge of veins. It is not very common and occurs only in some serpentine masses. The largest deposits of asbestos are in Canada, e.g. Black Lake and Thetford, and these locations produce approximately two-thirds of the world total. Other important deposits are located in Shabanie, Zimbabwe; Transvaal and Barberton, South Africa; USA; and Mt Fibro, Australia. In Europe it is worked especially in Cyprus; Bazenovo, Urals, USSR; and Dobšiná, Czechoslovakia.

Palygorskite (basic hydrous aluminosilicate with magnesium)
Monoclinic; H. 1—2; Sp.gr. 2.3

Sepiolite (basic hydrous silicate of magnesium)
Orthorhombic; H. 2—2.5; Sp.gr. 2

Both these secondary minerals originate by the decomposition of magnesium minerals. Palygorskite occurs in carbonate veins, and is found in hydrothermal deposits, e.g. Příbram, Czechoslovakia; Schneeberg, Austria; and with minerals of the so-called Alpine paragenesis. Sepiolite is formed by the weathering of serpentine, and is often associated with magnesite and opal. The most important deposits are Eski Shehir, Turkey; Bosna, Yugoslavia; Vallecas, Spain; and sites in the USA.

3

Chrysotile forms parallel fibres in serpentine (1). The fibres are very fine (0.01—0.06 mm), up to 30 cm long, and their tensile strength is between 40 and 80 kg/mm². It is a greenish, greyish, or yellowish mineral of silky lustre. (The name chrysotile comes from the Greek *chrysotil*, golden fibre.) It is widely used in the production of fire-resisting materials, filters, and insulating materials. Palygorskite occurs in flat or undulating leathery coatings ('rock leather') of white-grey, yellowish, or brownish colour (2). Sepiolite ('Sea foam') forms porous nodules with reniform surfaces (3). It is white and grey-yellow, and of feeble lustre. It is used in the manufacture of pipes and decorative objects.

2

Feldspars (aluminosilicates with potassium, sodium, and calcium)

From the mineralogical as well as the practical point of view, feldspars form the most important group of silicates. They are basic to the petrological classification of magmatic rocks, are mined as important minerals, and in the course of their decomposition provide substances which may be essential in plant nutrition as well as important raw materials for the ceramic industry. According to their chemical composition they may be divided into potassium feldpars (orthoclase) and sodium-calcium feldspars (plagioclases). The most common potassium feldspars are orthoclase and microcline, and their varieties; plagioclases include albite, oligoclase, andesine, bytownite, labradorite, and anorthite.

Orthoclase $KAlSi_3O_8$
Monoclinic; H. 6; Sp.gr. 2.53—2.56

Orthoclase is a rock-forming mineral found in acidic igneous rocks and some gneisses. As a secondary transported mineral it forms arkoses (sedimentary rocks composed of feldspar and quartz). It originates from magmatic rocks and hydrothermal solutions, and as an alteration product of other minerals under high temperatures and great pressures. By hydrothermal decomposition it readily changes into muscovite (sericite), zeolites, and other minerals. It easily becomes affected by weathering, and under certain climatic conditions changes into kaolin. Large accumulations of orthoclase formed in pegmatites are worked as raw material for the manufacture of porcelain, enamels, and in glazing pottery. The main supplier of feldspars is the USA (up to 60 per cent), but large deposits are also mined in Canada and Japan. In Europe, the most important deposits include some in Norway and Sweden. Druses of perfect crystals come from San Piero, Elba; Baveno, Italy; Strzegom, Poland; and Manebach, FRG.

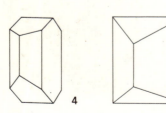

4   7

Orthoclase forms columnar (4) or tabular (5) crystals up to tens of centimetres long. It appears embedded or attached, and is often twinned (e.g. Carlsbad Twins) (1, 6). It also forms grains and coarsely granular aggregates. It is perfectly cleavable, fragile, breaks with splintery fracture, and is opaque or translucent. It is flesh-coloured, reddish or greenish, and of glassy or pearly lustre.

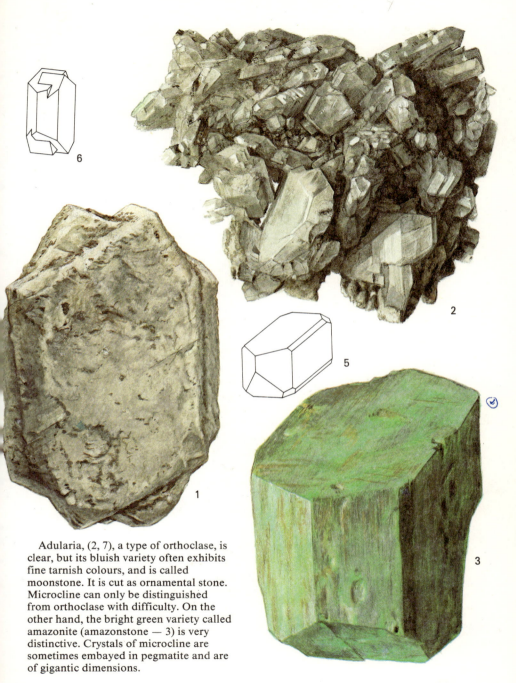

Adularia, (2, 7), a type of orthoclase, is clear, but its bluish variety often exhibits fine tarnish colours, and is called moonstone. It is cut as ornamental stone. Microcline can only be distinguished from orthoclase with difficulty. On the other hand, the bright green variety called amazonite (amazonstone — 3) is very distinctive. Crystals of microcline are sometimes embayed in pegmatite and are of gigantic dimensions.

Feldspars (continued)

One of the orthoclase varieties is the clear, transparent adularia, which is of hydrothermal origin. It occurs in many deposits of minerals of the Alpine paragenesis, and less often in ore veins (e.g. Baia Sprie, Romania). Tabular sanidine (from the Greek *sanis*, plate) originated in trachytes and phonolites. Examples are found at Drachenfels near Bonn and in the vicinity of Laacher See, FRG, and also in volcanic ejecta on Vesuvius. It is illustrated on the previous plate.

Microcline $KAlSi_3O_8$
Triclinic; H. 6; Sp.gr. about 2.5

It is found in large quantities in pegmatites, where individual crystals reach a weight of many tonnes. Its green variety, called amazonite (amazonstone), forms crystals up to 40 cm long. The most important deposits of amazonite include Iveland, Norway; Hagendorf, FRG; Mias, Southern Urals, USSR; and Pikes Peak, Colorado, and Delaware, Pennsylvania, USA.

Albite $NaAlSi_3O_8$
Triclinic; H. 6—6.5; Sp.gr. about 2.62

Albite occurs in basic igneous rocks, pegmatites, volcanic rocks, and gneisses. It originates from hot solutions in fissures and ore veins, as a rock constituent of minerals of the Alpine paragenesis. Fine druses are found in Schmirn, Austria; Scopi, Switzerland; Strzegom, Poland; and Andreasberg, FRG.

Labradorite (calcium-sodium plagioclase)
Triclinic; H. 6; Sp.gr. 2.7

Labradorite is an important constituent of basic igneous rocks (norites and gabbros) and some gneisses. Crystallized labradorite is found on Etna. It shows iridescence, and in polished form is used in wall tiling and in the manufacture of various decorative objects. It is mined at St Paul, Labrador, Canada; Ojamo, Finland; and Kosoj Brod, Ukraine, USSR.

3

In the deposit at Devil's Hole, Beryl Mine, Colorado, USA, a monocrystal of microcline was found which measured $49 \times 36 \times 13.5$ m, and weighed 15,900 tonnes; this is the largest known crystal. Albite crystals are columnar or tabular (1, 3), often with twin striation (pericline). Grains are embedded in magmatic rocks or gneisses. Albite is usually colourless or whitish, and is sometimes coated with chlorite. It is perfectly cleavable, glassy or clouded, and translucent or transparent. The play of colour in labradorite (2) with a bluish-green to yellow-orange sheen is most probably caused by the reflection of light from microinclusions of magnetite, ilmenite or haematite in cleavage planes, the bluish tinge being the interference colour of light refracted on microcracks in cleavage planes.

2

1

Zeolites (hydrated aluminosilicates with calcium, sodium, potassium, barium and strontium)

Zeolites are widely distributed minerals of similar chemical composition, appearance, and physical properties, arising in similar mineralization processes. Their name is derived from the Greek *zeo*, I bubble (boil), and *lithos*, stone. When heated by a blowpipe, they lose their water content and bubble and boil ('boiling stones'). They lose water easily and can regain it. This ability of zeolites to interchange water molecules by ions and other liquid or gaseous substances is of great technical importance. They are used as highly efficient chemical filters to prevent air pollution, and as organic fertilizers in agriculture.

Heulandite $CaAl_2Si_6O_{16}$. $5H_2O$
Monoclinic; H. 3.5—4; Sp.gr. 2.2

Heulandite usually crystallizes in vesicles in melaphyres, basalts and agate cores, e.g. in Oberstein, FRG, and Kozákov, Czechoslovakia, and in ore veins. It is very common in all parts of the world.

Thomsonite $NaCa_2Al_6Si_6O_{20}$. $6H_2O$
Orthorhombic; H. 5—5.2; Sp.gr. 2.3

Thomsonite occurs in vesicles in basaltic rocks and phonolites, and rarely in pegmatites; often it is the product of decomposition of nepheline. It is found, for instance, in the Central Bohemian Highlands, Czechoslovakia; perfect crystals come from New Jersey, USA.

Natrolite $Na_2Al_2Si_2O_{10}$. $2H_2O$
Orthorhombic; H. 5—5.2; Sp.gr. 2.2—2.4

Natrolite is the most common zeolite. It is found in vesicles in young volcanic rocks. It was described for the first time in 1824 based on material from Mariánská Hora, Czechoslovakia. Slender prismatic crystals of a length of many tens of centimetres are found in occurrences in the USSR, and in Poonah, Anguilla, and St Berthelemy, India.

2

1

Heulandite crystals are thin- to thick-tabular (1, 2) or columnar; aggregates are usually scaly (lamellar). Heulandite is brittle, perfectly cleavable, of pearly lustre, colourless, whitish or sometimes brick-red, and transparent or translucent. Thomsonite crystals have the form of longitudinally striated columns terminated with a base (3), but frequently they form reniform or globular aggregates (4) with an orthorhombic parquet-like surface. Thomsonite cleaves perfectly, has glassy or pearly lustre, and is white, yellowish, greenish, reddish or colourless. Acicular to capillary crystals of natrolite are often terminated with pyramids (5) or form bunches (6). Aggregates are fibrous and silky. It is a colourless and white, glassy mineral.

3

4

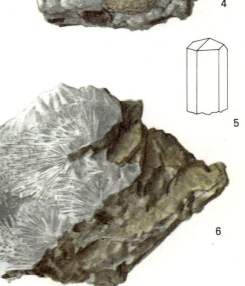

5

6

215

Zeolites originated partly in the final phase of the solidification of basic magmatic rocks; most of them, however, formed during hydrothermal processes. Zeolites fill vesicles in young volcanic rocks from which gases have escaped, and they are also formed in pegmatites. Less commonly, they also crystallize in hydrothermal ore veins, in deposits of magnetite or native copper, or are deposited at hot springs. They are widely distributed in volcanic areas all over the world. Important deposits are found in India. Zeolites are also formed secondarily, e.g. by the decomposition of plagioclases and nepheline, filling cracks in different rocks.

Stilbite (desmine) $CaAl_2Si_7O_{18} . 7H_2O$
Monoclinic; H. 3.5—4; Sp.gr. 2.1—2.2

Stilbite is widely found in cavities and fissures in volcanic rocks, forms druses in pegmatites, e.g. in Strzegom, Poland, occurs associated with minerals of the Alpine paragenesis, and in ore veins.

Chabazite $CaAl_2Si_4O_{12} . 6H_2O$
Trigonal; H. 4.5; Sp.gr. 2.1

Chabazite, a common zeolite, is a typical volcanic mineral. It is also found, less frequently, in ore veins and fissures in granites. The best known occurrence in Europe is in Řepčice, Czechoslovakia.

Apophyllite $KCa_4F(SiO_5)_4 . 8H_2O$
Tetragonal; H. 4.5—5; Sp.gr. 2.3—2.4

Apophyllite occurs in cavities in volcanic rocks, in fissures in limestones, and in ore veins; it is also deposited at thermal springs, e.g. Plombières, France. Perfect, large crystals are found in New Jersey, USA; India; and Brazil.

Leucite $KAlSi_2O_6$
Tetragonal and cubic; H. 5.5—6; Sp.gr. 2.45—2.5

Leucite is often found in basic volcanic rocks. Its characteristic crystals are found on Vesuvius, Italy, and elsewhere. Large specimens come from São Paulo, Brazil, and from Magnet Cove, Arkansas, USA.

4

Stilbite crystals (2) form bundle-shaped bunches (1). Stilbite is cleavable, of pearly or greasy lustre, and colourless, honey-yellow, or reddish in colour.

Rhombohedral crystals of chabazite (3) resemble cubes (4) and are often intergrown. The lenticular to tabular variety is called phacolite. Chabazite is brittle, of uneven fracture, translucent or transparent, and of glassy lustre. It is usually colourless, white, yellowish, red, or brown. Apophyllite crystals are pyramidal, columnar (5), cubic or tabular (6). It cleaves perfectly, is brittle, of glassy or pearly lustre, and is transparent or translucent. Sometimes it is also milky dull, and it is whitish, pink, yellowish, rarely red, greenish, or bluish in colour. Leucite is the first mineral in which a large quantity of potassium was found. It occurs only crystallized in a typical form (7, 8). It does not cleave, is glassy, and is whitish, yellow-grey, or reddish in colour.

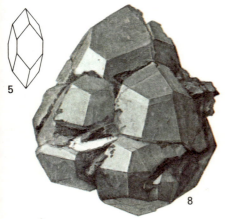

8

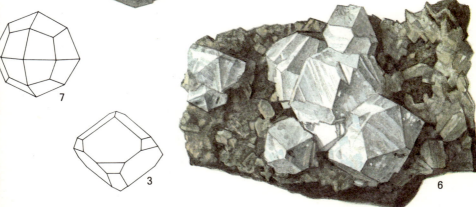

6

Whewellite $CaC_2O_4 . H_2O$
Monoclinic; H. 2.5; Sp.gr. 2.23

Whewellite, which is a crystallized organic compound, is neither common nor widely distributed in nature. Its sporadic occurrence in coal or ore deposits was known to miners of long ago; they took it, however, for calcite or barytes. It was included in the mineral system as late as 1840 under the name of whewellite, in honour of the English natural scientist W. Whewell.

The first specimens probably came from the ore deposit in Cavnic, Romania; later it was also found in Freiberg, GDR. Usually this calcium oxalate occurs in sedimentary rocks in coal and lignite basins. It was deposited here by mineralized solutions circulating through the sediments. Together with other minerals it crystallizes in particular in cracks in clay ironstones ('septarian nodules'), giving rise to a specific minerogenetic association. At the beginning of the 20th century, the most important occurrence of this type was the coal basin in Kladno, Czechoslovakia, where the most perfect many-faced crystals up to 5 cm long have been found. Other locations are the coal deposit at Burgk near Dresden and the coal basins near Zwickau, GDR; Most, Czechoslovakia; and Alsace, France. Large quantities of whewellite, especially in the form of aggregates, were discovered recently in Maikop, Caucasus, USSR, and elsewhere.

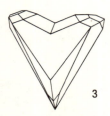

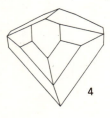

Whewellite crystals are mostly many-faced, tabular, or columnar (2), and are often twinned with arrow-shaped terminations. The mineral forms typical heart-shaped (3) and butterfly-like twins (4), clustered in bunches (1). Granular aggregates are usually composed of long fibres. Whewellite is colourless, limpid, milky-dull in aggregates, and of conchoidal fracture. Crystals are brittle, and have several perfect cleavages. The largest known whewellite crystal (5.5 × 5.5 × 1.5 cm) comes from Burgk, GDR. Associated minerals in clay ironstones include ankerite, barytes, calcite, millerite, sphalerite, and galena. Barytes and calcite are minerals which are similar to whewellite.

2

1

Amber (succinite)
(mixture of oxygenous hydrocarbon compounds)
Amorphous; H. 2.—2.5; Sp.gr. 1—1.1

Amber is the fossil resin from conifers growing in Tertiary times on a large area south of the Baltic Sea. Originally it was deposited in the so-called blue soils, from which it was gradually shifted to secondary deposits, i.e. river and sea alluvia. Some amber nodules reveal preserved insects and vegetable tissues, and the paleontological investigation of these organisms made it possible to determine the age of amber at approximately 50 million years. Resins looking very similar to amber are found in some brown coal basins or in older sediments, and are usually known under local names, e.g. neudorffite, retinite, siegburgite, trinkenite, and valchovite. They all originated in a similar way.

Amber is found in all countries located along the coast of the Baltic Sea. The most important deposit is at Jantarnyi, USSR, and many other localities are known in Poland and the GDR. Amber of inferior quality was found near Kiev and Lwóv, Ukraine, USSR, and in Romania. Dark brown amber, called simetite, comes from Sicily.

Amber was known to man as early as the Stone Age. In ancient times it was very much valued as a gemstone, known under the Greek name *elektrón*. It used to be transported from the Baltic Sea to Rome along the famous 'amber paths'.

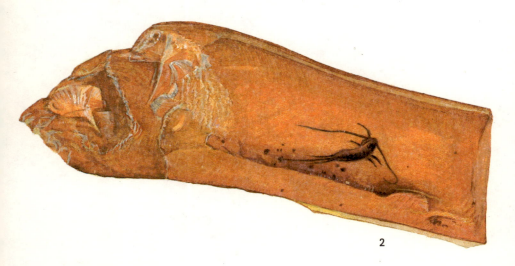

2

1

Amber is found in the form of grains, pebbles, or irregular nodules (1), exceptionally weighing up to 8 kg. It is compact, brittle, conchoidal in fracture, greasy, translucent or transparent. Its colour is honey-yellow, whitish, or reddish to brown, and it is often banded. On rough surfaces it is dull and opaque. The enclosed Tertiary insects (2) are perfectly preserved.

When rubbed with a piece of cloth amber becomes electrically charged and attracts light objects. This property of amber has been known since ancient times and in fact the word electron is derived from the Greek *elektrón*, meaning amber. As an ornamental stone amber was particularly valued in the 17th and 18th centuries. However, it has remained a popular component of jewellery up to the present time.

INDEX